contents

the speed of
nearly everything

tobogganing penguins
spinning neutron stars

the speed of
nearly everything
from tobogganing penguins
to spinning neutron stars
Peter Macinnis

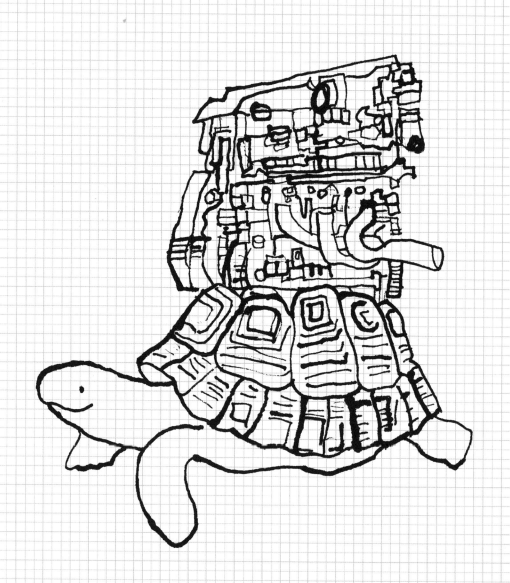

why speed facts & stats?

As a small boy, I wondered about the speeds of things – as small boys do – and marvelled at the tales of the cheetah and the botfly. I only became aware of other people's interest in really fast things in 2002, when the University of Queensland invited me to attend a test-firing from Woomera in the outback of Australia.

My excuse to make the trip out there was that I was writing a history of rocketry. I had just come back from the Aerojet plant in California, where they fill tubes 'five feet across and fifty feet long', roughly 1.5 metres by 15 metres, with high explosives. When you watch a rocket launch, the big boosters that provide most of the fireworks are these very same tubes. They are the big lumps of uncontrolled, unintelligent grunt that get the rest of the rocket moving before they are discarded and fall back to Earth.

The university engineers wanted to test something far more subtle, but to do so, they had to send a rocket 320 km (200 miles) up into the air so that when it reached its peak, it could turn over and plunge back to the ground, accelerating furiously. Just 6 seconds before it crashed into the semi-desert, a supersonic ramjet, or scramjet, would fire up, the first time a scramjet had been tested under real conditions in the atmosphere.

There are some limitations to these really fast things that will need to be solved if humans are to have a future in space. The booster rockets of a normal spacecraft, like the more intelligent liquid-fuelled rockets that fire as the booster tubes drop away, illustrate these limitations.

The fumes they give off are generally highly toxic, and they add to the greenhouse load. Worse still, the rockets we use now are inefficient because they need to carry both the fuel that they require, and an oxidiser that makes the fuel burn.

While gunpowder rockets burn carbon and sulfur, 60% of rocket powder is saltpetre, which is needed to make the carbon and sulfur burn – all rockets need to carry some type of oxidiser. Some use kerosene and a nitric acid oxidiser, but many are far grimmer or harder to handle – and even filthier in their output.

There are other options, though. Liquid hydrogen and liquid oxygen in pressurised tanks make a clean liquid-fuelled rocket, but both gases are very hard to handle and to contain. The rocket also needs eight times as much oxygen as hydrogen, and the oxygen has to be carried aloft for the sole purpose of burning the hydrogen.

Jet aircraft don't have this problem. Although they fly at high altitudes where the air is far thinner than it is at ground level, the engines can still grab the oxygen they need for combustion from the air outside. The scientists wondered if they could do the same for rockets, at least until the rockets are far enough up to leave the atmosphere behind.

Enter the scramjet, a supersonic ramjet. While it is something of an oversimplification, think of a funnel that scoops up air as the craft rushes along, and you will have a rough idea of what the scramjet does. The reality is a lot more complicated. The engine only operates at Mach 4 or above (that is at least four times the speed of sound), but it could be very efficient.

Designs for yachts, prop-driven aircraft and other vehicles that go at less than the speed of sound can be tested in a wind tunnel. Even the German V2 rockets could be tested (up to a point) in wind tunnels. At speeds close to or above the speed of sound though, it gets harder and harder to achieve a non-turbulent wind – a laminar flow. As the speed gets higher, the conditions in the wind tunnel become less and less like the reality of flight through a still atmosphere.

The University of Queensland engineers had been testing their models in what they called a 'shock tunnel'. The T4 shock tunnel uses hot gas under extremely high pressure, which is rammed until it bursts through a 3 mm ($^1/_8$ inch), sheet of steel that the engineers call 'a diaphragm', to howl past whatever needs to be tested.

The problem, they told me, was that they could not be sure how close the results they got were to reality. So they decided to traipse across Australia to bolt their scramjet onto a rocket and launch their dreams, and their hopes, into space.

The Woomera landscape is a surreal one. It is red and dry; hot by day, even in the southern winter, and cold when the night is clear, even in summer. Plants are scattered across the sandy surface, with stones in between. One early and rather foolish English explorer thought the plants and stones would help stop the water evaporating when it rained, making the area excellent land for grazing stock.

Kangaroos and emus live there, and there has always been a nomadic Aboriginal presence, but after World War II, Britain decided it would also be a great place to test rockets. Aircraft can be warned away, and if something goes up 320 km (200 miles), like the scramjet was to do, well into space, the launch must be timed to make sure that no satellites are passing overhead.

There was a window, not of opportunity so much as inopportunity that morning for our rocket as a satellite went by, but the launch went off without a hitch. We amateurs winced as the white rocket exhaust spiralled, but the professionals assured us that the flight was clean, the spiralling caused by wind shear acting on the exhaust.

Then there was the nervous wait while the rocket climbed steadily and turned over to fall back to Earth, thanks to forces known to the ancient Greeks as early as 500 BC and detailed by Isaac Newton in 1666. The rocket fell, and fell, far downrange, where telemetry vehicles waited to gather the data from the last 6 seconds. Back at the launch site, the tower caught a few scraps of data before the scramjet fell below the Earth's curvature, and it all looked good. We spoke of the Wright Brothers' 12 second first flight at Kittyhawk in 1903, and Goddard's first rocket flight in 1926, which lasted 3 seconds. A 6-second flight seemed enough for a pioneering effort, given the magnificent speed.

By the time we knew the data matched the shock tunnel measurements, I had driven at Mach 0.1 back to Adelaide and flown, somewhat closer to the speed of sound, back to Sydney to start my write-up. As giant road trains hurtled past me in the pre-dawn light, also going at Mach 0.1, but in the other direction, it struck me just how excited ordinary people were about speed.

President Reagan's speechwriters once had him declare that scramjet aircraft would allow people to fly from Paris to Tokyo in 2 hours, and the cliché is still trotted out when scramjets are mentioned; 'Sydney to London' or 'New York to Moscow', or some other long haul, and always in 2 hours.

Forget that – scramjet passenger flights would be hellishly expensive and you wouldn't even have time to open that second bottle of free champagne before landing. They fly extremely fast, and while they may, one day, do the heavy work of lifting us into orbit, they're not there just yet. Their true destiny is as fast workhorses, not racehorses. Don't spread the word though, because we humans do love speed!

Scientific abbreviations

Centimetre per second	cm/s	Litre	L
Feet	ft	Metre per minute	m/min
Feet per second	ft/s	Metre per second	m/s
Foot pound	ft lb	Metre	m
Frames per second	fps	Miles per hour	mph
G-force	g	Millimetre	mm
Hertz	Hz	Nautical miles	nmi
Joules	J	Pound	lb
Kilogram	kg	Revolutions per minute	rpm
Kilometre	km	Second	s
Kilometre per hour	km/h	Square metre	m^2
Kilometre per second	km/s	Yards per second	yd/s

A note about units and precision: some of the values I quote are precise measures and some are approximations. Most of my readers will be familiar with either Imperial or metric (SI) units, while a rare few will know both systems. The last group will soon see that I have generally (but not always) opted for consistent precision in the conversions, even when this introduces spurious precision. Communication counts for more, so I have rounded a few conversions where the precision was confusing. I hope the sophisticated minority will absolve me of this peccadillo. Also, I have not distinguished between short and long tons.

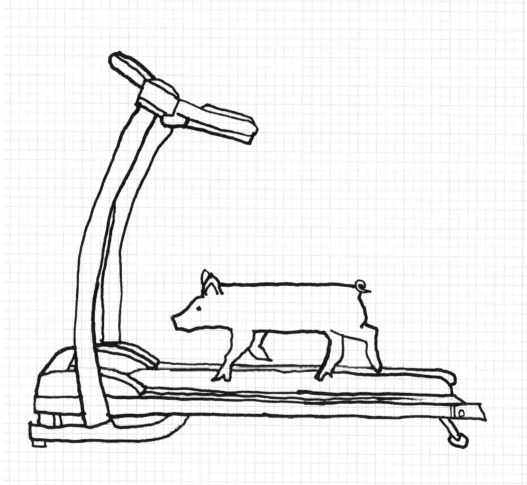

running with the pack

Snakes don't just slither – some of them fly (or glide, anyhow) and others swim. And while most fish swim, flying fish take to the air and mudskippers walk. Mammals, on the other hand, swim, fly, hop, leap, bound, crawl, amble and run – and many of them move pretty fast. Insects are every bit as versatile – even some snails are fast. Plants don't move, though some of their seeds do, and even the animals that settle down, like oysters and barnacles, have at least a few days as free-swimming animals.

Some of the animal kingdom live slow and steady lives, savouring each moment, while others prefer the fast and furious. Humans are quite unusual as they live long lives at breakneck speeds.

From top to bottom, from the poles to the equator, animals are moving. A Rüppell's vulture collided with an aircraft in 1973 at 11,277 m (37,000 ft), and a few years earlier, whooper swans were sighted at 8200 m (26,900 ft). Submersibles have also found life forms at the very bottom of the oceans, all moving freely.

As animals move, they need muscles and nerves to control them, and they need lots of energy to keep those muscles working and the nerves sparking over. And whoever doesn't move fast enough is destined to appear on the menu of some other animal. A bite of animal contains a great deal more energy than a mouthful of vegetation, so chasing other animals to eat is a good investment for most animals, and avoiding being eaten is always a good strategy.

why animals move fast

Moving helps animals find or catch food – moving fast helps some animals avoid being caught as food. The fastest runners are usually either good food or serious hunters.

All animals must have some basic resources, which is why so many animals show territorial behaviour, chasing off interlopers. They have to move around to get to the water, food and shelter they need, and they must move fast enough to stop some other intruder from moving in and taking some essential need away from them.

The Arctic tern for example flies from the Arctic to the Antarctic each year, and in doing so lives in something close to a perpetual summer. This means they fly close to 19,000 km (12,000 miles) twice a year. Migratory animals like the tern travel these long distances each year to avoid the extremes of climate, and many of them need to travel fast just to be sure of some rest and feeding time at each end.

The Arctic tern breeds in the northern hemisphere. In 1982, to track a tern's movements, a chick was banded in its nest in Britain. Just three months after fledging, it was already in southern Australia, near Melbourne, having flown down the coast of Africa, across the Indian Ocean, and across much of Australia, a grand total of 22,000 km (13,500 miles). That means it had travelled 245 km (155 miles) a day. Terns keep this up for life, flying an estimated 800,000 km (500,000 miles) in a lifetime.

Salmon need to swim fast as well, to get upstream past the rapids and waterfalls. They can reach speeds of up to 8 m/s (26 ft/s) or 30 km/h (18.5 mph) in still water, and they use this speed to get up through rushing waters to reach the headwaters where they will breed and die – if they are lucky.

Hummingbirds on the other hand may appear to move slowly or even hover in one place at times, but careful high-speed photography reveals that they beat their wings between 75 and 90 times a second, combining slow and fast in one operation. And they can keep it up for quite some time – the ruby-throated hummingbird (*Archilochus colubris*) can fly for up to 30 hours, covering 800 km (500 miles) non-stop from Rockport, Texas to the Yucatan Peninsula in Mexico.

No matter what the obvious reason may be, in the end, most animals move fast because it helps their survival. Only humans seem to travel fast just for the fun of it. Some scientists even think that we are human *because* we run … but we will come to that later.

the animal clocking problem

There is a difference between judging how 'fast' something is, and counting the length of units covered in a single time unit. And it isn't always easy to time a twisting, swerving animal – few animals run in a straight line so it is difficult to pinpoint just how fast they really are. A fox can run faster than a rabbit, but as the rabbit is better at dodging obstacles, the fox has to follow along, becoming harder to 'clock'. Many animals rely on short bursts, while others manage on endurance alone.

Some birds can be timed using radars, and there are a few birds that can be timed by getting them to follow an ultralight aircraft. Certain birds, geese among them, seem to be hardwired to follow the first moving thing they see after hatching from their egg. In nature, this thing is usually their mother – this imprinting lets them learn to recognise and seek out mates from among their own species.

Imprinting can be a problem when birds are being reared in captivity and imprint on their carers, later rejecting biologically suitable partners, preferring humans instead. If an experimenter allows a gosling or other newly hatched bird to imprint on him or her, and the experimenter later flies off in an ultralight, the imprinted adult bird will fly along as well.

It is then a simple matter of raising the speed of the aircraft until the bird cannot quite keep up, and reading the air-speed indicator.

Horses and dogs can usually be persuaded to run in straight lines, so they can be reliably timed. And if you're brave enough, charging elephants can also be clocked by driving away just fast enough that you're not in danger of being flattened, while somebody else checks the speedometer. Many quoted animal speeds are just a little bit rubbery though, and those for the botfly and cheetah are plain dubious.

Sometimes, we can figure out an animal's speed with physics. For example, if a salmon leaps up a 6 m (19.6 ft) waterfall, we know that it took off at the same speed you would have if you fell vertically through 6 metres. In Figure 14, Non-terminal velocities, page 159, this is calculated to be around 39 km/h (24.4 mph).

doppler radar

A Doppler radar is useful in figuring out how fast things are moving – it can be used to measure the speed of birds, tennis serves or a golf drive, the speed of an aircraft, the advance of a weather front, or in speed traps for cars. It all depends on something that Christian Doppler noticed in 1842.

To put it simply, the Doppler effect is what you experience when a train goes through a level crossing, or when a car passes you with a siren or horn blaring: the frequency of a tone or of any other wave signal depends on the relative speeds of the source and the detector. If the source and the detector are getting closer, the sound we hear is higher in pitch; if they are getting further apart, the pitch is lower.

The bell on the level crossing sounds a deeper note after we pass it, as does the siren or horn as it moves away – and with the right instrument, we can measure the difference and estimate a speed. That is what the Doppler radar does. It compares the frequency emitted and the frequency that is bounced back, and gives a measure of how fast the reflecting object is travelling.

The most distant galaxies are moving away from us very fast, and this shows up in the red-shift of their light as light frequencies are lowered. There are lines that appear in all spectra at fixed places and these also shift: if these lines appear to have moved to a different place in the spectrum, we know that the source is moving.

If the object is travelling across the radar beam, it appears to have no velocity at all. Somewhere in that nugget of knowledge, there is a way of beating radar speed traps, but I haven't quite worked it out yet. Sadly, the police are already working on a system which will be able to measure vehicle speeds from side-on. Just as I was getting somewhere!

the speedy botfly that wasn't

When you start to look into fast animals, more often than not, you will read that the fastest animal of all is the deer botfly, credited with an amazing 1287 km/h. If you convert this figure into miles per hour, it comes out as a round 800 mph, a figure that smells a little bit of fudged science – and rightly so.

The story begins with a 1927 article in the *Journal of the New York Entomological Society* by the entomologist Charles Henry Tyler Townsend, who reported a similar speed. He claimed that the botfly was clipping along at 366 m/s (400 yd/s), which is 1316 km/h (818 mph). As we will see, any precision in this conversion is hardly justified.

Townsend reasoned that as these flies passed in a blur, they must have been travelling very fast. On that scientific basis, and no other, he credited them with a nice round 366 m/s (400 yd/s).

This story should have been questioned right away, but people wrote it down, passed it on, quoted it uncritically, and never stopped to wonder what would happen if botflies were really tearing around at supersonic speeds.

As we see later, while some people took the time to prove that bumblebees couldn't fly, nobody stopped to consider or demonstrate the impossibility of the botfly claim until 1938, when Irving Langmuir, a Nobel laureate in chemistry, tested the assumptions.

Firstly, the air pressure on the botfly at that speed would be more than half an atmosphere, surely enough to crush it. The energy needed to maintain a flight at that speed would be 370 watts – half a horsepower – which would be quite an ask. Aside from anything else, travelling that fast, the botfly would use up its own weight in fuel every second, so it would need to be a very fast feeder.

Next, Langmuir had been hit by these botflies. While it hurt, that weight of botfly at 1316 km/h (881 mph) would have left a significant hole, rather like that of a soft bullet, and the botfly would have been mashed inside the wound. Instead, the botflies simply bounced off him.

Langmuir mocked-up a model of the botfly, using solder to make a 1 cm ($^2/_5$ inch) long, 0.5 cm ($^1/_5$ inch) wide pellet. He attached this to a string, and whirled it around his head, timing it so he could work out its velocity. He reported that at 21 km/h (13 mph) it was a blur, at 42 km/h (26 mph) it was barely visible, at 69 km/h (43 mph) an observer could not tell which way it was going, and at 103 km/h (64 mph), it was completely invisible.

He concluded that the blur Townsend had seen came from a botfly travelling at 40 km/h (25 mph). His results were published in *Science* and reported in *Time* magazine, but legends are tough things, even when debunked by Nobel Prize winners. So even today, the same old botfly values keep emerging from the woodwork.

By a curious chance, Langmuir's name crept into the record books in an entirely different way in 2006 when plasma physicists used a specially designed holographic-strobe camera to capture pictures of matter waves travelling at 99.997% of the speed of light.

Known as Langmuir waves, they are generated by intense laser pulses, and may one day lead to 'tabletop' versions of high-energy particle accelerators. One of the steps along the way was to take photographs of the waves to see if they behaved like scientists thought they would. They did.

the cheating cheetah record

The champion running hunter is undoubtedly the cheetah, but, as in so many of these cases, the figures have been a little, well, bent out of shape. The cheetah legend began with an article by Kurt Severin in *Outdoor Life* magazine in 1957, based on the observations of a trained cheetah named Ocala. Severin and Ocala's trainer marked out a 73 m (80 yd) track and set up a hand-cranked lure and a meat-scented bag; the cheetah was encouraged to run this bag down, and a shot was fired when the big cat reached the other end. The time was recorded as 2.25 seconds, which was converted to 114 km/h (71 mph).

Although this figure is still commonly quoted, by 1959 critics had noted that the track used was really only 59 m (65 yd) long and that if the time was taken as correct, the calculated speed should have been 117 km/h (72.7 mph). On the 59 m (65 yd) track, that time would give a speed of 95 km/h (59 mph). There always remained a risk though that the timing had been unreliable, because of errors in human reaction time, and so others began their own investigations.

Using some Disney film shot at 64 fps, Milton Hildebrand estimated that the cheetah in that film ran at a respectable 90 km/h (56 mph). Then, in 1965, Craig Sharp, a veterinary surgeon in Kenya, got access to a tame cheetah. He measured out an accurate 201 m (220 yd) course, and used a Land Rover to trail meat for the cheetah to chase.

The animal was allowed to have a running start (Severin's Ocala had started from a crouch), and the same operator clicked the stopwatch at both the start and end of the run. Over three runs, times of 7.0, 6.9 and 7.2 seconds were recorded. These equate to speeds between 101 and 105 km/h (62 and 65 mph), with 102 km/h (63 mph) a reasonable compromise. *That* is the figure which should appear in the record books.

penguin pace

Penguins don't fly, but they do have a few other methods of travel. They hop from place to place, they toboggan when they can, they walk and they swim underwater, though if you have ever seen penguins swim, you might say they were 'flying in water'.

Hopping penguins do not usually go in a straight line, so we have no data for the speeds of hopping penguins, but they do manage to reach a reasonable speed for short bursts.

Very few people have ever tried to time a tobogganing penguin, but observers in the Antarctic report that the penguins 'go faster than a running man'. The point of tobogganing to a penguin though is not that it is fast: it is useful simply because it is more energy efficient when compared with walking. An emperor penguin walks at 2.8 km/h (1.7 mph), and the Adélie penguin waddles at 3.9 km/h (around 2.4 mph).

A swimming jackass penguin has been timed at 7 km/h (a bit over 4 mph), while chinstrap penguins have been clocked in the water at 24 km/h (around 15 mph), the same speed as a good middle-distance human runner on land. The smallest of the penguins, the Little Penguin of Australia, has been reported as swimming at 5.5 km/h (3.4 mph) but this figure is for sustained swimming in a straight line. Gentoo penguins are pretty fast, having been credited with swimming at speeds of 40 km/h (25 mph). This seems high, but they often dive to 100 m (328 ft), with dives taking as little as half a minute, averaging out at 24 km/h (15 mph). The calculated speed would be quicker still if we allowed for the time in the dive for them to catch and/or swallow fish along the way.

migratory birds

Over the years, many scientists have spent time studying the performances of long-distance migratory birds. A recent research paper combined a number of sources, from which the following examples have been drawn. All values are for flapping flight, not gliding flight or while diving, and the speeds were all measured by tracking radar. The figures here have been somewhat simplified from the original source, an article on flight speeds among bird species.

There is one puzzling feature about some of these observations though: the Caspian tern covers a mere 40 to 50 km (25 to 30 miles) a day when it is migrating, a distance it could quite easily cover in about 1 hour of flapping flight. This could perhaps be because it devotes a fair amount of time to gathering food. By contrast, the ruby-throated hummingbird can fly 800 km (500 miles) nonstop in just 30 hours.

Figure 1. The speed of birds

Scientific name	Common name	Speed (m/s)	Speed (km/h)	Speed (mph)
Botaurus stellaris	Great bittern	8.8	32	20
Hirundo rupestris	Crag martin	9.9	36	22
Falco naumanni	Lesser kestrel	11.3	41	25
Accipiter nisus	Sparrow hawk	11.3	41	25
Larus minutus	Little gull	11.5	42	26
Larus ridibundus	Black-headed gull	11.9	43	27
Sterna caspia	Caspian tern	12.1	44	27
Merops apiaster	European bee-eater	12.2	44	27
Larus argentatus	Herring gull	12.8	46	29
Turdus pilaris	Fieldfare	13.0	47	29
Larus canus	Common or mew Gull	13.4	48	30
Parus major	Great tit	13.6	49	31
Corvus corax	Common raven	14.3	51	32
Phoenicopterus ruber	Greater flamingo	15.2	55	34
Pelecanus onocrotalus	White pelican	15.6	56	35
Sturnus vulgaris	Common starling	16.2	58	36
Branta canadensis	Canada goose	16.7	60	38
Cygnus cygnus	Whooper swan	17.3	62	39
Anas platyrhynchos	Mottled duck	18.5	67	42
Gavia stellata	Red-throated diver	18.6	67	42
Anas crecca	Common teal	19.7	71	44

Scientific name	Common name	Speed (m/s)	Speed (km/h)	Speed (mph)
Anas acuta	Northern pintail	20.6	74	46
Anas penelope	Eurasian widgeon	20.6	74	46
Aythya fuligula	Tufted duck	21.1	76	47
Aythya marila	Greater scaup	21.3	77	48
Polysticta stelleri	Steller's eider	21.9	79	49
Clangula hyemalis	Long-tailed duck	22.0	79	49
Melanita nigra	Common scoter	22.1	80	50
Aythya ferina	Pochard	23.6	85	53

a peloton of pelicans

Migrating birds like ducks, cranes and geese traditionally fly in a V formation or a 'skein', with one bird leading and two 'wings' trailing out behind – and so do some pelican species. Because birds can feel the effort that they put in to flying with every beat of their wings, even the young ones learn that there is a lot less effort and a great benefit to come out of flying this way.

The V formation does not help birds to fly faster in terms of distance per hour, but it helps them to fly further each day for a given amount of effort. The mathematics and the science get quite complicated, but in simple terms, each bird generates an eddy, called a wingtip vortex, and the work required to start the air spinning works as a drag, holding the bird back.

In technical terms, each wingtip vortex causes a downwash that increases the induced drag. The downwash for one bird is an upwash for the bird following behind, assuming it is in the right position: behind and slightly above the bird in front. The second bird gets free lift, which means it does not need to flap its wings as hard to generate the thrust it needs to move forward while keeping its altitude. In this situation, the muscles don't work as hard, reducing the heart rate, giving the birds more energy for flying. In theory, a flock of 25 birds can fly up to 70% further than 25 solo birds could fly with the same energy. But does it work that way in practice?

Dr Henri Weimerskirch and his colleagues were able to test this in 2002, using great white pelicans (*Pelecanus onocrotalus*) which had been trained to follow a motor boat or an ultralight aircraft. They fitted the birds with tiny heart rate monitors, and then got them to follow along at various speeds, either solo or in formation. The birds' heart rates dropped when they were flying as a group and they were able to glide more often.

So what is in it for the lead bird? It has to work the hardest of the whole flock, but each following bird helps to reduce the downwash of the bird in front, reducing the drag for that bird as well. The best position is to be somewhere in the middle of the line, with at least one bird in front, and one behind.

In short, there are three hard-working positions; at the head of the V, and in each tail position. They may not know any science, but the birds can tell the difference, so when the lead bird begins to get tired, it will drop back down the V and another bird will take over. In the same way, the birds in the trailing positions will move up the line, leaving other birds to take over. It seems that over time, each bird will share the load.

Different species have different habits. Fisherman's Wharf tourists in San Francisco at dusk often see small groups of brown pelicans, *Pelecanus occidentalis*, rushing to join existing formations. In sunny Australia, the Australian pelican, *Pelecanus conspicillatus*, makes a lot of use of 'thermals' – hot air currents – that they can ride to high altitude (sometimes as much as 3000 m, or 10,000 ft) before gliding off alone to the next thermal at speeds as high as 55 km/h (34 mph). If you can get a free solar lift, why hang around waiting for other pelicans?

To these birds, speed is less important than energy, but there is another aspect to formation flight. When birds are migrating, formation flying probably helps prevent losing young birds. The Australian pelicans can usually see nearby thermal columns and the wheeling pelicans riding them, so there is less chance of birds getting lost, and no need for tight formations.

the art of outrunning a crocodile

According to the experts, Australian crocodiles are not always out to eat you. If they feel threatened, they will sometimes bite defensively and then retreat – if they do hang on though, they are looking at you as a possible dinner.

On the other hand, if a crocodile bites and then lets go, but stays where it is, it is planning a better grip, so run! People-eating saltwater crocodiles (*Crocodylus porosus*) can reach speeds of 10 or 11 km/h (6 or 7 mph) when they 'belly run'. They are often faster if they are sliding down muddy tidal riverbanks – the equivalent of a crocodilian toboggan (with teeth), but that is downhill on mud.

Uphill and on dry land, the saltwater crocodile, has no chance of catching an active human. In water, it is another matter. The 'salties' can swim at an estimated 10 or even 15 km/h (6 or even 9.5 mph), though they rely mainly on being able to surge out of hiding to grab their prey.

If a crocodile doesn't let go, you need to remain calm and fight back. If you can, go for the eyes or the nostrils – admittedly, if your hand is already inside its jaws, this does limit your choices. The experts say you should reach down for the palatal valve, which stops water running into their throats. Opening this floods their lungs, causing them to drown – leaving them unable to continue the attack.

I'd like to shake the remaining hand of whoever came up with that one!

Luckily for Australians, the saltwater crocodile cannot 'gallop' like the smaller and less aggressive freshwater crocodile. The 'freshies' can reach an impressive 17 km/h (10.5 mph), but they tire after 20 to 30 m (65 to 100 ft). Crocodilians found in New Guinea, Cuba and along the Nile are also able to gallop.

outrunning a dinosaur

The top speed of a human running 100 m (109.5 yd) is about 10.2 m/s (33.5 ft/s), a little over 36 km/h, or slightly less than 23 mph. If you are an elite athlete, you might be able to run a 4-minute mile, at about 15 mph (24 km/h). So how would you go against a dinosaur?

In short, be glad the dinosaurs are dead. A careful scientific analysis undertaken by two scientists in 2007 looked at running speeds in a variety of birds to get estimates of dinosaur speeds which revealed that we would stand little chance.

You might be able to escape *Tyrannosaurus rex*, with a bit of luck, because *T. rex* could only make 8 m/s (26 ft/s), about 29 km/h (18 mph). More importantly, it would probably be a bit slow to get started and might not have been all that good at turning – but any slip would be fatal for you. An *Allosaurus* would have been more dangerous: it was probably able to run at 9.4 m/s (31 ft/s), 34 km/h (21 mph), while *Dilophosaurus* could scoot along at 10.5 m/s (34.5 ft/s), 38 km/h (more than 23 mph).

Compsognathus was the most dangerous of all, probably making 17.8 m/s (58 ft/s), around 64 km/h (40 mph).

Even the horror of Jurassic Park, the *Velociraptor*, could turn in an Olympic performance, hitting 10.8 m/s (35.5 ft/s), 39 km/h (just over 24 mph). About the only good news is that *Velociraptor* has just been reinterpreted. New evidence says that it had feathers and may have been more like a big turkey, though it probably still would not have been a turkey to mess with!

outrunning big birds

Even if the *Velociraptor* was more of a big turkey, some larger modern birds, the flightless ones known as ratites, present quite a threat. Because they are so large, the rhea, the emu, the ostrich and the cassowary are unable to fly up into the trees to escape, so they are fast running, and have a nasty kick that most mules would die for – and they all run faster than we do.

The exception to the rule is the New Zealand kiwi, which had a fellow ratite, the giant moa, until less than a thousand years back. Perhaps the kiwi found a safe niche, away from any size overlap with moas or their young.

The ostrich of Africa is far larger than the Australian emu, weighing around 65 kg (143 lb), an average weight for a human adult, while emus weigh less than half that at just 27 kg (37 lb). The emu has a slightly longer stride, but each pace takes more time, possibly because the emu's leg is 9 cm (4 inches) shorter, than the ostrich's legs are. The combination of these two factors sees the emu reach a creditable 13.3 m/s (43.5 ft/s), 48 km/h (30 mph), but the ostrich takes the gold with a top speed of 15.4 m/s (50.5 ft/s), 55 km/h (34 mph).

These two birds can be persuaded to run in a straight line over open country, so they can be timed fairly easily from a motor vehicle. The cassowary of northern Australia and New Guinea is a jungle bird with no enemies, so it rarely needs to hurry. Anecdotal evidence says they can chase prey at 30 km/h (18 mph), while the South American rheas are generally somewhere between the emu and the ostrich. Kiwis don't run very fast, but they are small and good at hiding. The moa could almost certainly have outrun a human, like the other big ratites, but judging by its absence on the scene today, it seems not to have run fast enough in the past.

animals we can all outrun

The slow loris mainly eats slow food – large snails, insects, lizards, small mammals, eggs and fruits – so for the most part, it does not need a fast chase. Even so, it belies its name and strikes at its food very fast indeed by standing, gripping a branch with both hind feet, and then hurling itself at its prey.

The three-toed sloth has even less need to travel fast. Its green colour helps it merge into the vegetation as it moves along the underside of branches, living up to its name. On the ground, an adult sloth can travel at 2 m/min (6.6 ft/min). In the trees, it speeds up to 3 m/min (9.8 ft/min), or, if it is answering a distress call from its young, a female sloth may reach 4 m/min (13 ft/min).

There is a natural speed limit for birds because if they fly too slowly, they stall and fall out of the air. The slowest speed that a bird can safely manage appears to have been achieved by the American woodcock (*Scolopax minor*) and the Eurasian woodcock (*Scolopax rusticola*). These birds have both been timed flying at 8 km/h (5 mph) without stalling during courtship displays.

On the other hand, where American or New World hummingbirds can buzz around at 75 to 90 wing beats a second, several New World vultures have been reported to beat their wings just once a second in level flight.

The slowest animals of them all are the 'sessile' ones. They attach themselves mainly to rocks and stay in one place – this choice does make demands on them though. Firstly, they have to get their food from their environment, and that means they must live underwater, as there is little food floating in the air. Next, they need to spread out, so they begin their lives in a free-swimming or drifting stage, having to find a suitable place to attach to at some point, usually after a few days.

longevity in animals

Some animals can live for a very long time, but few of them outlast human beings. Even in the nineteenth century, when life expectancy in Europe, the United States and Australia was less than 40, anybody who got to adulthood had a reasonable chance of living to 80. The world average life expectancy is now 67, according to the World Bank.

Sometimes, a short (and hopefully happy) life is all that is available. The males of a mouse-like Australian marsupial, *Antechinus stuartii*, stop producing sperm cells when they are 10 months old, and then indulge in a frantic race to mate as often as possible before they die of exhaustion about 2 weeks later. The mating sessions last from 5 to 14 hours.

The right sort of longer life must *surely* be better than that! A white whale called Mocha Dick (as the name suggests, the inspiration for Herman Melville's *Moby Dick*) was attacking whaling ships as early as 1810. He was finally killed by a whaler in 1859, when he was seen to show signs of old age. He had worn-down teeth, he was blind in one eye, and he must have been close to 60 by then.

Figure 2. Life expectancy of common animals

Animal	Age life expectancy (years)	Record age (years)	Animal	Age life expectancy (years)	Record age (years)
Mouse	1–3	4	Dog	10–12	24
Golden hamster	2	8	Wolf	10–12	16
Rat	3	5	Pig	10	22
Guinea pig	3	6	Lion	10	29
Groundhog	4–9	n.a.	Deer	10–15	26
Kangaroo	4–6	23	Sheep	12	16
Rabbit	6–8	15	Goat	12	17
Chicken	7–8	14	Monkey	12–15	29
Budgerigar	8	12+	Bear	15–30	47
Fox	8–10	14	Ass	18–20	63
Cow	9–12	39	Horse	20–25	50+
Pigeon	10–12	39	Hippopotamus	30	49+
Duck	10	15	Elephant	30–40	71
Cat	10–12	26+	Human	67	120

human longevity explained

So why do humans last so long? An anonymous doggerel writer apparently thought that liquor intake played a role in it all. In some branches of biology, ethanol is in fact used as an excellent preservative. Unfortunately, it only works when it is used on dead tissues, so this logic, often heard on licensed premises, rests on a false premise.

The horse and mule live thirty years
And nothing know of wines and beers;
The goat and sheep at twenty die
And never taste of Scotch or rye;
The cow drinks water by the ton
And at eighteen is mostly done;
The dog at fifteen cashes in
Without the aid of rum or gin;
The cat in milk and water soaks
And then in twelve short years it croaks;
The modest, sober, bone-dry hen
Lays eggs for nogs, then dies at ten.
All animals are strictly dry,
They blameless live and early die.
But sinful, ginful, rum-soaked men
Survive for three-score years and ten –
And some of us, a very few,
Stay pickled till we're ninety-two.

– Anon.

development times in animals

How quickly can a living thing grow from a single cell, a fertilised egg? It does depend on the species, but there are always a few extra things to consider when coming up with an answer. With birds for example, we look at the time from laying to hatching, and for other animals, we take from conception to birth. That said, the gestation time shown in Figure 3 for the kangaroo needs some further explanation.

Figure 3. Gestation periods of common animals

Animal	Gestation or incubation (days)	Animal	Gestation or incubation (days)
Pigeon	11–19	Wolf	60–63
Golden hamster	15–17	Pig	101–130
Budgerigar	17–20	Lion	105–113
Mouse	19–31	Goat	136–160
Duck	21–35	Monkey	139–270
Rat	21	Sheep	144–152
Chicken	22	Bear	180–240
Rabbit	30–35	Deer	197–300
Groundhog	31–32	Hippopotamus	220–255
Kangaroo	32–39	Human	253–303
Emu	49–54	Cow	280
Fox	51–63	Horse	329–345
Cat	52–69	Donkey	365
Dog	53–71	Whale	365–547
Guinea pig	58–75	Elephant	510–730

Female kangaroos can store a fertilised ovum for some time before allowing it to develop. Also, the anatomy of female marsupials makes a large birth canal impossible (the problem is complex, but best understood as a plumbing problem), so all marsupials are born through a narrow canal in a very undeveloped state. They then attach to a teat in the mother's pouch, where they effectively remain in gestation for a considerably longer time. It all contributes to making the figure less useful in comparisons.

Humans are also a little misleading: the head and brain are far larger than for other animals, and human babies continue to develop for a long time after they are born. No other young animal is so helpless for so long after it is born, and we should probably add at least six months when we compare ourselves with other animals.

fast and slow fish

The slowest fish is believed to be the seahorse, which sits in the water in a vertical position, swimming by gently rippling its pectoral fins. Estimates suggest that they would not cover more than 150 m (164 yd) in an hour. To swim faster, a fish needs to use its tail and its body to drive itself through the water.

The speed that a fish can achieve depends on its size and build, but even fast fish usually have two speeds: a cruising speed and a full attack/flight speed. The species that hunt other fish must, logically, be the fastest, as you can see in Figure 4.

Sharks are the great hunters among fish, but little information exists on their speeds. The mako shark is said to reach 100 km/h (62 mph), but this is hard to confirm. They do travel long distances though. A single tagged great white shark called Nicole travelled more than 20,000 km (12,400 miles) from South Africa to Australia, and back again, in just 9 months. Nicole travelled at a minimum speed of 4.7 km/h (2.9 mph) during this journey.

Figure 4. The speed of fish

Fish type	km/h	m/s	mph
Herring	6	1.67	3.7
Pike	6	1.67	3.7
Carp	6	1.67	3.7
Cod	8	2.22	5
Mackerel	11	3.06	6.8
Salmon	45	12.5	28
Bonito	60	16.7	37

Fish type	km/h	m/s	mph
Small tuna	60	16.7	37
Bonefish	64	17.9	40
Swordfish	64	17.9	40
Wahoo	78	21.7	49
Black tuna	80	22.2	50
Marlin	80	22.3	50
Sailfish	110	30.5	68

even plankton and crinoids

Plankton got their name because they supposedly never swim. Officially, they drift in currents – *planktos* is a Greek word meaning 'drifting'. While the tiny plants of the sea, the phytoplankton, may indeed drift, recent research has shown that the tiny animals we call the zooplankton swim actively.

Without plankton, nothing could live in the sea. The phytoplankton, tiny single-celled plants, are eaten by the zooplankton, the tiny animals that are food for the smallest fish – and for some of the largest animals, like some of the whales.

So how do you study microscopic animals and their behaviour? The answer came in the form of acoustic imaging. Scientists illuminated the animals using 1.6 megahertz pings of ultrasound, and this showed them that the plankton can 'treadmill', keeping their depth constant against upwellings and downwellings, and doing it at a significant speed.

Crinoids are echinoderms, close relatives of feather stars and more distant relatives of seastars and sea urchins, but even if they have close relatives in the animal kingdom, their common name, 'sea lily', reveals what most biologists make of these animals. They have a stalk that attaches to rocks and long feathery arms that wave gracefully in the currents, and they seem to be more like plants.

Videos taken at 430 m (470 yd) in the Bahamas in 2005 revealed something surprising: sea lilies creeping along the sea floor, apparently to escape the attentions of predatory sea urchins. The sea lilies do anchor on the sea floor, but they have the ability to break their stalks above the attachment point and just below a set of finger-like appendages that they can use to attach to a new, safe place.

speedy sea mammals

The information relating to the speed of marine mammals is patchy at best. The dolphins which play in front of a speeding vessel may seem to be swimming as fast as the ship that follows them, but they are actually riding on a wave of pressure that is being pushed along by the ship itself. The common dolphin (*Delphinus delphis*) has been timed at 60 km/h (37 mph) while riding a ship's bow waves, but only 45 km/h (28 mph) in open water. Dall's porpoise (*Phocoenoides dalli*) has been reported as reaching 56 km/h (35 mph).

Baleen whales cruise the ocean at around 1 m/s (3 ft/s), 3.6 km/h (2 mph), but one of the baleen whales, the blue whale, the largest living animal species, cruises closer to 20 km/h (12 mph). When alarmed, blue whales can increase this to 50 km/h (31 mph). In the era before the fast, modern whaling ships, this speed helped the blue whales escape the whalers' harpoons.

35

Blue whales can leap completely out of the water, which probably involves raising their centre of gravity by 8 to 10 m (26 to 33 ft). It is easy to show that their speed when they fall back into the water is equal to their speed as they started to emerge, and a 9 m (30 ft) fall generates a speed of almost 50 km/h (31 mph), as shown in Figure 14, Non-terminal velocities, page 159.

The sei whale (*Balaenoptera borealis*) can easily hit 55 km/h (35 mph) in short bursts, while the killer whale (*Orcinus orca*) can keep that speed up for much longer. The fastest seal known is the California sea lion (*Zalophus californianus*), which has been recorded at 40 km/h (25 mph).

Female whales are fast milk deliverers. The calf simply has to hang on as its mother pumps milk at the rate of 4.5 L (1 gallon) per second. Calves that don't hang on get a faceful of milk.

the running vampire bats

Bats can fly brilliantly, but at a cost. Like birds, their wings are their front limbs, but unlike birds, bats are unable to fold their wings neatly out of the way. Part of the problem is that mammals have no feathers, and so, bats need a greater area for their wings, which are covered with thin membranes, and that means bigger wings. Whatever the cause, bats do not usually perform well on the ground.

The exception to that rule is the vampire bat, which is not a myth. Unlike the fictional vampire of Gothic novels though, they do not transmit vampirism – but they do feast on blood in their home range, which is in Central and South America. *Desmodus rotundus* not only walks though, it also runs! There is only one other walking bat, a native of New Zealand, but it cannot run.

Vampire bats feed by landing a small distance from their prey, usually large mammals, sleeping cattle for example. They sneak up and bite a small nick, usually in the animal's heel, and then lap up the blood. The wound continues to flow for some time, because the bat's saliva contains an anticoagulant which prevents clotting.

A scientist called Daniel Riskin was studying the bats' behaviour and became interested in seeing how fast they could go, so he put a vampire bat on a treadmill. It made no real sense to do this, as bats can't run, but nobody told the bat, so it ran.

The vampire bats run rather like gorillas, using their strong front limbs (their wings) to propel themselves along. On the treadmill, they were able to run at 4 km/h (2.5 mph), but that still doesn't answer the question of why they can run. Riskin thinks that perhaps the bats used to prey on smaller, more agile animals, and may have needed to be able to run if their prey decided to lash out at them. The research continues.

turtles and tortoises

Legend has it that in 456 BC, the Greek playwright Aeschylus was killed when a bird dropped a tortoise on his bald head, mistaking it for a rock. The bird is usually called an eagle, but a less noble-sounding bird, a Lammergeier vulture, probably did him in. These sagacious carrion-eaters are in the habit of flying high with large bones and dropping them to get to the marrow inside – apparently they think of tortoises as slow-moving bones, suitable for dropping.

The tortoise is celebrated in ancient myth for its slowness. The hare and the tortoise contest, however, taught generations that slow and steady wins the race. That did not stop Pheidippides from running himself to death after the battle of Marathon ('The Marathon', page 57).

Animals need speed to get to food before it escapes, or to avoid being caught and eaten. For the most part, land-based tortoises are herbivores, though the occasional insect may be munched by chance. In general, tortoises need not hurry for dinner, and with their shells, they are reasonably well protected from predators, so speed is unimportant.

A giant tortoise has been timed at 5 m/min (16.5 ft/min), or about 0.17 mph (0.27 km/h), but it is unlikely that it would keep it up for any length of time. On land, sea turtles are lucky to reach 2 km/h (1.2 mph), but in the sea, it is quite different. A green turtle cruises somewhere between 14 and 18 km/h (9 and 11 mph) in the sea, and a Pacific leatherback has been reported at 35 km/h (22 mph). All marine turtles are capable of considerable speed when threatened. Immature green turtles are carnivorous – leatherbacks eat jellyfish and other soft marine life-forms – and some sharks can chomp through turtle shells with ease, so marine turtles need a good turn of speed.

the riddle of the chameleon's tongue

' Why, sometimes I've believed as many as six impossible things before breakfast.'
The White Queen, *Alice's Adventures in Wonderland* (1865) by Lewis Carroll

Working out the speed of an animal or how it travels often involves working out limits. As an undergraduate, I once dissected the flight muscle of a pigeon, clamped a hook to each end, and then hung the muscle from a beam, dangling a bucket from the lower hook and pouring water into the bucket until the muscle broke.

By measuring the breaking strain of a pigeon chuck steak, you can find an upper limit to how hard the pigeon can beat its wings. Along with wind-tunnel tests on a pigeon wing, film of pigeons flying and other data, you can come closer to understanding how a pigeon flies, from a scientific viewpoint. Pigeons have a much smaller brain than us, but they understand precisely how they fly. The problem is that they cannot explain it.

As we will see in the strange case of the bumblebee (see 'The art of estimation', page 235), scientific tests like this often deliver a contradiction, suggesting that what we can see happening in front of us is impossible. Clearly it is *not* impossible: the contradiction just means that one or more of our assumptions must be wrong, and we need to look into it further.

The chameleon is an all-round impossible animal at first glance. Consider the way it changes colour to blend in (but cuttlefish can do that as well), the way it swivels its eyes (but some snails, crabs and flies also have eyes on stalks), and then consider its tongue. However you look at it, the chameleon's tongue is *totally* impossible.

The chameleon is one of the more unusual members of the lizard family, catching its prey with a powerful, whip-like tongue that flashes out as far as one-and-a-half body lengths, grabs the food and whisks it back to its mouth. In the Netherlands in 2004, Jurriaan H. de Groot and Johan L. van Leeuwen used high-speed video and X-ray film to assess how it all happens.

First up, the acceleration is tremendous. In 20 milliseconds, the tongue is travelling at 6 m/s (20 ft/s), producing a g-force of 51 *g*, and given what we know about power production in muscles, this was *completely* impossible. In other words, some important detail was missing from the picture.

Stating the problem in less emotive terms, it was clear that muscle alone could not be responsible for the acceleration. This is not a new problem, but none of the previous solutions have been all that satisfactory, most of them requiring blood pressure to throw the chameleon's tongue out, a bit like one of those party favours.

The answer is that the tongue operates on the principle of a catapult or a bow. We could never throw an arrow or a stone as fast as we can fire them from a bow or a slingshot, yet our muscles perform the 'impossible' by storing energy in the elastic of the slingshot or bow, and then releasing it very quickly.

The chameleon's tongue has elastic collagen connecting the tongue bone and the accelerator muscle. The collagen is arranged in a series of concentric sheaths that fly outwards when they are released, sending the adhesive tongue tip flying at the prey.

However, the impossible chameleon has nothing on a tropical ant of Central and South America (*Odontomachus bauri*), which hurls itself 8 cm (3¼ inches) into the air and up to 40 cm (16 inches) sideways, just by snapping its jaws shut at 65 m/s (213 ft/s). The snapping action can also knock potential prey unconscious. Now *that* is worth believing before breakfast!

lizards and snakes

Most reptiles move fairly slowly, because they are what we call 'cold-blooded'. In reality, this means that their blood is the same temperature as their surroundings, and can actually be even higher than our standard 37°C (98.6°F). Most biochemical reactions go much faster after a slight increase in temperature, so when you read reports of reptilian speed, you also need to know the temperature.

After the chameleon's tongue, one of the most amazing reptilian speed shows is the run of the 'racehorse goanna' of inland Australia. The word 'goanna' is a corruption of iguana, but there is little similarity between iguanas and the monitor lizards, a group which includes both the goannas and the Komodo dragon of Indonesia.

When the temperature is high, the racehorse goanna can hit 25 km/h (15.5 mph), which is bad news for the other Australian lizards that are on the goanna's menu. The lashtail lizard gets up on its hind legs and hits 22 km/h (13.6 mph), usually enough to get it to shelter, but the slower ornate crevice dragon manages by staying close to a rock crevice and accelerating fast back into it when danger looms. Around the world, lizards face problems like this and solve them with speed or cunning.

Sometimes, they are just lucky to have the right ancestors. This certainly applies to the monitors, which have a major advantage: they can run and breathe at the same time. At some stage in their evolution, the more advanced animals needed to develop a way to pump air into their lungs as a way of improving their aerobic capacity. Their ability to obtain oxygen helps them to 'burn' food faster to make energy.

There are various ways of pumping air: we humans have a diaphragm between our intestines and our lungs, a sheet of muscle that we can pull down, forcing air into our lungs. Kangaroos use the thump of landing to clear their lungs, and the stretch of the next hop to draw fresh air into the lungs.

Some lizards twist their bodies from side to side as they move their feet forward, but they also use the same muscles to rotate their ribs and expand their lungs, pulling air in – there is no easy way of using the same muscles for two tasks, and that means most lizards quickly run out of breath.

Monitors use a 'gular pump', a set of throat sacs to pump air into their lungs as they run, but how effective is it? Tomasz Owerkowicz knew how to find out: he put a bunch of lizards on the treadmill, three green iguanas and six monitors, and then took X-ray videos of them as they ran. Later, he blocked the gular pumping, and the monitors' performance fell to that of the iguanas.

Monitors have another big advantage. They have a much larger lung surface area and a heart with four chambers, like mammals, rather than the simple and inefficient three-chambered hearts found in other reptiles. It seems the racehorse goannas and other monitors chose their ancestors wisely!

Among snakes, the African black mamba probably takes the lead – it is variously credited with 12 to 20 km/h (8 to 12 mph). Snakes generally rely on being able to strike fast, within their own length, though even then, they usually only move at 1 to 2.5 m (3 to 8 ft) a second, relying on closing the gap before their victim has time to react.

Most snakes travel comparatively slowly when they are moving across the ground – luckily for us, black mambas use their speed mainly to avoid threats. All the same, their speed is enough for the mamba to be blamed in African folklore as the cause of whirlwinds.

exit, pursued by a bear

That stage direction in Shakespeare's *The Winter's Tale* is the last we hear of Antigonus, who perishes off-stage, saving on the special-effects budget.

The number of words of Czech that I know could be counted on the fingers of a clumsy saw-miller's hand. I do, however, know that to the Czechs, a bear is a *medved*. I know that because a Czech who is a bit of a philologist told me about the word and how it was derived. The literal meaning is 'honey-knower', and my friend pointed out that the *med-* part appears in the English word *mead*, while the *-ved* can be seen in the Sanskrit *Vedas* – and in *wit*.

So we have one word, spanning much of the world, a bit like the bears, who are a variable bunch. I would have thought 'honey-knower' might be a better label for the industrious bees: it takes twelve worker bees their entire lives to make just one tablespoon of honey. Instead, we give the name to bears, some of whom know where to find honey.

The name 'grizzly bear' is older than the dictionaries believe: Thomas Malory's *Morte d'Arthur*, written in 1470, makes reference to a 'gresly bear', and the expression clearly migrated to America. Given the choice between being pursued by a grizzly bear and a polar bear, there isn't very much in it.

A polar bear lopes along at 40 km/h (25 mph) and a grizzly bear is good for 45 to 50 km/h (28 to 31 mph). In either case, if pursued, you are likely to be caught. Polar bears don't have much experience with trees, but if one appeared while you were being chased, you would need to climb fast, as they can jump 2 m (6 ft) in the air. They can also swim at 10 km/h (6 mph), faster than the fastest Olympic swimmers. Regrettably, young grizzlies and some adults can climb trees.

running cats and dogs

Fox-hunters are accused of claiming that their sport is harmless, and that the fox actually enjoys it. I find it hard to believe that even the unspeakable in pursuit of the uneatable would be that silly, but nonetheless, hunting by cats and dogs is good for a prey species, so maybe foxes benefit? It is bad for individual animals, but good for the herd, as the old, the sick and the unfit are eliminated, leaving more food for the rest. Such is the harsh calculus of population biology.

Hunting also makes predators fitter. Like many of the hunting cats, the cheetah is a sprinter, and can only run at its top speed for about 450 m (490 yd). It can still keep up to 72 km/h (45 mph) after that, but only for about 4 to 5 minutes before it slows down.

The lion is even more limited, but relies on a combination of an 80 km/h (50 mph) charge and the support of the rest of the pride to bring prey down. Leopards and jaguars are credited with 65 km/h (45 mph), the puma has been recorded at 50 km/h (31 mph), and most of the big cats have similar speeds. Even the domestic cat can deliver a sprint of 45 km/h (28 mph).

The fastest domestic dog seems to be the Afghan hound, which has been timed at 70 km/h (43 mph), though coyotes have been seen doing almost the same speed over short runs. The Cape hunting dog has been clocked at 72.5 km/h (45 mph), the greyhound at 62 km/h (38.5 mph).

Over 200 m (219 yd), the whippet can achieve 57 km/h (38 mph), the grey fox can achieve 42 km/h (26 mph) – and climb trees as well! – while the red fox can reach 50 km/h (31 mph) when being chased, and it can swim, giving it another avenue of escape. The hyena is able to notch up 64 km/h (40 mph) when it needs to, while the jackal manages 55 km/h (34 mph).

when the deer and the antelope run

There is an old shaggy dog story about two hunters who are being chased by a bear, a wolf, or some other fast animal. One stops to change into running shoes, the other gasps that the shoes will not help him outrun the bear, to which the first man replies 'No, but they will let me outrun you!'

Most evolution is about shading the odds. Just because you are 'more fit', there is no guarantee that you will survive, because luck always plays a part. In a game where the house percentage is extinction for the species, everything else comes down to the odds. Over time, predators get faster, and if there are fast predators around, the fastest prey will escape more often.

But then, if the predator happens to become extinct (perhaps because it can't catch its dinner any more), you are left with a special ability or a characteristic that has no great use. The North American pronghorn antelope for example, runs at almost 100 km/h (62 mph), close to the top speed of a cheetah. This is hardly surprising, given that cheetahs (*Acinonyx trumani*) were in the southwestern United States until 10,000 years ago.

Other animals, similar to today's African lions, elephants and camels, were there as well as the cheetah. Some conservationists argue seriously that the old Pleistocene balance should be restored by introducing equivalent modern African species into the United States.

Animals near the top of the food chain, the top-order predators, help shape ecological systems and maintain biodiversity. When the United States lost most of its wolves and grizzlies, elk numbers increased. Elks ate willows, a favourite beaver food, so beaver numbers dropped 80–90%. Fewer beavers meant fewer beaver dams and fewer wetlands to support the willows, so willows are now down to 40% of what they were when wolves and grizzlies roamed the land.

boing, lurch, waddle and crump

In the Australian bush, marsupial mammals are, in varying ways and depending on your point of view, quite lazy or supremely efficient.

By day, kangaroos rest in the shade of a tree, hidden among long grass, waiting for the darker hours to feed. If they are alarmed, they hop off very fast. A human trying to hop would find this extremely tiring, but all kangaroo and wallaby species are built for hopping, and for speed. Red kangaroos cruise at 20 km/h (12 mph), and in full flight they can reach 65 km/h (40 mph), leaping as high as 3 m (10 ft) with 8 m (26 ft) gaps between hops.

The most obvious thing about a kangaroo is its huge tail, which it uses as a counter-balance. As mentioned earlier, each hop pumps air in and out of the lungs. That, along with the highly efficient way energy is stored on each leap, gives the kangaroos and wallabies their edge.

Other marsupials in Australia are fast as well – in his poetic study of bushfire, *Death of a Wombat*, Ivan Smith said the common wombat went 'waddle and crump'. The smaller hairy-nosed wombat can sprint at up to 40 km/h (25 mph) for 90 seconds or so, usually enough to reach the shelter of a burrow.

The koala, another popular Australian local, is not a bear at all, but actually a wombat relative that climbs trees. Wombats burrow and the females have a backward-opening pouch to keep dirt out as they dig. This is also seen in koalas, which live in trees, where the pouch is hard to explain, except as an evolutionary leftover. They spend as much time as they can there, coming down only to change trees. Koalas rarely amble more than 200 m (650 ft) in the night, unless they have been removed from their home range. They are probably slowed down by the toxic oils in the *Eucalyptus* leaves, but it is these same oils that make koala meat unpleasant. With flesh like that, who needs speed?

fast insects

Ants usually travel at about 6 cm/s (2 inches/s), while ravaging army ants step up the pace to about 8 cm/s (3 inches/s), when the path is clear. If the way is not clear, Scott Powell and Nigel Franks found that army ants engage in an odd behaviour. The researchers forced ants to use a timber bridge with 'potholes' bored in it, and showed that some of the ants will actually climb into the holes and grip the edges, making a smoother road for other ants – now that's a commitment to speed!

Ordinary bees can fly as fast as 7.25 m/s (24 ft/s), but at this speed they lose their rotational stability, and are likely to roll over. To avoid this, they lower their legs to stabilise their flight by producing lift. This also works by reducing the moment of inertia, rather like a skater who extends his or her arms.

Cockroaches are fast, very fast – just ask anybody who has met one in the kitchen late at night! Or at least that's how it seems. Careful tests show that they actually only move at human walking pace, 5.41 km/h (3.36 mph). Their comparatively small size and the fear they arouse in most people just make them seen faster.

The fastest known insect is the sphinx moth, at 53 km/h (33 mph), followed by the botfly (maybe!), and an Australian dragonfly, clocked at 36 km/h (22 mph). Butterflies and moths have set a few other records over the years: some butterflies have been sighted at a height of 6000 m (20,000 ft) and North African painted lady butterflies have winged it 6400 km (4000 miles) to Iceland! They must have either been blown fast, flown fast, or have amazing stamina.

tiger, tiger, running blind

Until the late 1990s, entomologists were puzzled by the odd behaviour of tiger beetles. These insect predators (with more than 2000 species around the world) hunt down and eat other beetles, grasshoppers, ants and caterpillars, but as they hunt their prey, tiger beetles will often stop dead before restarting the chase.

Cornell University entomologist Cole Gilbert discovered in 1997 that the tiger beetle goes blind when it accelerates towards its prey. Not for long, but long enough for them to have to engage in these stop-and-start attacks. As he explained it, beetles simply do not get enough light in their eyes to form an image, and they have to stop and take a sight on their prey again.

This works for the tiger beetle, because it runs very fast, and can make up the lost time. Gilbert found that a 10 mm ($^1/_3$ inch) American beetle could rush along at 0.5387 m/s (1.9 km/h or 1.2 mph). Another species from Australia, twice that length, travels at 2.5 m/s (9 km/h, or 5.6 mph).

As he sees it, this means that the Australian beetle runs 125 body lengths per second, while top sprinters cover about 5.5 body lengths per second. On that basis, the American beetle is ten times as fast as a human, and the Australian beetle is almost 23 times as fast. Still, at least human sprinters can see where they are going.

To date, the reason why the beetles lose track of their prey is unresolved. At the time, Professor Gilbert suggested that what works for the tiger beetle may also work for robots, so resolving the visual tracking systems that the beetles use might have applications on the Mars Rover or similar exploration robots on other planets. These robots need to move quickly to explore large areas, and that means a trade-off that could see the robots go blind as well.

snails and slugs

Snails and slugs don't like straight lines, so timing them is a challenge, and racing them frustrating. Even so, snail-racing events occur; one famous race happens each July in Congham, in England's Norfolk.

The way round this lack of snail discipline is to put them at the centre of a circle, usually described as '13 to 14 inches in diameter' (about

34 cm). The first snail to cross the perimeter is declared the winner. The record over this ill-defined course is held by a snail named Archie, who reached the edge in 2 minutes, in 1995.

On that basis, a determined and fiercely resolute snail might travel 10 m in an hour, 1 km in just over 4 days, and easily complete a seven-day mile. Ignoring this, snail racers cruelly start each race with 'Ready, Steady, Slow!' One banana slug took 2 hours to cross the perimeter, so the 1-year-mile record for banana slugs is still up for grabs.

Other molluscs can be surprisingly fast. Canny gardeners in Australia will know the flat-shelled snail, an entirely different species, has the magnificent habit of preying on the introduced garden snails, striking them in the snail-equivalent of the neck. Few believe the speed with which the carnivorous snails strike until they have seen one in action.

The deadly venomous cone shells are known to have killed at least thirty humans. The Conidae are sea snails found in tropical waters, especially around Australia. These hunting snails strike their prey with enough speed and force to pierce wetsuits. Even though cone shells crawl at 1.5 m/h (5 ft/h), while other snails go ten times as fast, their speedy strike makes up for that. In some corners of the molluscan world, shouting 'Ready, Steady, Slow!' is both unkind and unwise. *Never* taunt a cone shell!

horse race at the speed of light

A quarter horse is named for its speed over the quarter-mile (400 m). A good quarter horse averages 76 km/h (47.5 mph), though at the finish line, the horse will probably be doing close to 80 km/h (50 mph). At least over short distances, the zebra and the Mongolian wild ass are also fast, both reported to reach 65 km/h (40 mph).

The Melbourne Cup is Australia's premier horse race. It is run over 3200 m (close enough to 2 miles), and the fastest on record was 3 minutes, 16.3 seconds in 1990 by Kingston Rule. That is equivalent to an average speed of 58.7 km/h (36.5 mph).

The Grand National in Britain had its fastest time in 1990 when Mr Frisk completed the course in 8 minutes and 47.8 seconds. It is run over 4 miles and 4 furlongs (7242 m), and the race throws in the added difficulty of requiring the horses to jump 30 fences. The average speed for Mr Frisk's record run was 49.4 km/h (30.7 mph).

Secretariat won the Kentucky Derby in 1973 in 1 minute and 59.4 seconds. It takes place over 2 km (1.25 miles). This was 60.64 km/h (37.69 mph). Unusually, Secretariat ran each successive quarter mile faster than the previous one, and if allowed to continue, may well have exceeded the speed of light by about now. Luckily, it was reined in, saving the physicists a lot of embarrassment.

the pace of the pachyderms

The classification of animals and plants always involves a bit of opinion, sometimes causing confusion. While Aristotle did not exactly see that whales and porpoises were mammals, he knew they were not like fish. Linnaeus on the other hand, who invented our classification system, listed whales and porpoises as fish in the first ten editions of his book *Systema Naturae*.

Nineteenth-century scientists caused a similar problem when pulling elephants, rhinos and hippos into an illegitimate group, the pachyderms. They were all big, had thick grey skins, and came from Africa, but the grouping logic made as much sense as grouping worms and wombats because they burrow, or butterflies and birds because they fly.

Still, pachyderms were big and they had a formidable approach to threats: they charged them down. They were quite heavy enough not to fear anybody or anything. They still are, but they are still not related.

A rhinoceros, to start with, will charge for short distances at 40 to 50 km/h (25 to 31 mph), as timed by chargees in motor vehicles. While black rhinos (really just dark grey) have poor vision, and often break off, or run into a tree, they *are* very good at changing direction, which tends to take all the fun out of being charged. They can also be aggressive towards each other, and may keep up their charging speed for some time when chasing other black rhinos.

Hippos can certainly outrun a human on land, though estimates of their actual top speed vary between 30 and 50 km/h (18 and 31 mph). Hippos are vegetarians, but that does not seem to stop them from attacking and killing humans: they have a reputation for killing more people in Africa than lions, though the Cape buffalo is a contestant for that honour as well. There is some good news though: hippos can't jump!

how to tell when an elephant is joking

Elephants, the largest of the three pachyderms, walk at a sedate 7 km/h (4.5 mph), and they can keep it up for a considerable time. They have large territories and need to keep moving so as not to eat one area out. When it comes to fighting humans, their main enemy, they can accelerate to a higher pace.

African elephants will sometimes engage in what is called a mock charge, but at other times, they are deadly serious. In either case, the elephant will approach, people say, at some 50 km/h (31 mph). Reversing at this speed can be risky, so safari drivers need to know

51

which charge is which when 6 tons of elephant is heading their way. A word to the wise: in a mock charge, the elephant's ears stand out wide from the head and the trunk is curled; in a serious charge, the elephant has his or her ears back and trunk down – there is more to the charge than that though.

Researchers have discovered that elephants 'hear' through their feet, sending out rumbles at 20 Hz, so low that humans can hardly hear them. Sound travels through soil at around 3300 m/s (10,000 ft/s) (that's around 12,000 km/h, close to 6800 mph), almost 10 times as fast as in air. The low sounds travel amazing distances, as much as 10 km (6 miles). Given that the speed of elephant sounds through the ground is 30% of the escape velocity of our planet, it is just as well that elephants can't run that fast!

In nature, female elephants use the mock charge to chase off lions or hyenas, and the effect of moving the ears away from the head is to make her look even larger than she is. It is possible that the charge sounds emitted and transmitted across the African plains also vary, but that only other elephants can tell the difference.

There is just one problem with the safari-driver claims, and that is the speed attributed to the elephant: John Hutchinson (an American biologist who is now at the Royal Veterinary College, London) and his colleagues studied and videotaped large numbers of elephants, and found the highest speed observed was more like 25 km/h (15 mph).

And interestingly, the elephants don't actually run, even at their top speed, not according to Olympic standards. The official definition of a walk is that at least one foot must be on the ground at any one time, and while elephants have been snapped with three feet off the ground, they have never been caught lifting all four at once.

the bulls of pamplona and other stampedes

Each year, locals from the Spanish town of Pamplona and more than a few tourists join in 'the running of the bulls', part of the festival of Sanfermines. If you have your head about you, it is easy enough to do, because the bulls only need to travel 848.8 m (2785 ft) from the point where they are started, with the average time to get from A to B about 3 minutes and 55 seconds.

Note that this is the *average* time. These bulls are young, know no fear, and weigh 600 kg (about 1300 lb) each. In 1959, one bull completely lost interest in running, and took more than 30 minutes to cross the finish line, which would have made a bit of a hole in the average. Some of the bulls must run faster to make up for it.

Then there is the human side of the running of the bulls. There are 2000 runners a day on weekdays, and as many as 3500 on weekends, and at least some of those lining up know that each year about 300 people are injured, around 3% of them seriously. That means there is a lot of nervousness (or to be honest, panic) in the hearts of those waiting to run, or running. It also means that there are a lot of people who will be happy to trample on you, in order to avoid being trampled by some bulls.

So while the average speed of the bulls is only a bit over 13 km/h (less than 10 mph), no faster than a house mouse, there will be a few of them that will have heard about the chap who drowned in a lake with an average depth of 15 cm (6 inches). Besides, house mice are lighter, have less staying power, and don't travel in herds.

Without the other runners, Pamplona would be far safer than running in front of a herd of buffalo, especially American buffalo who can reach 50 km/h (31 mph). Make that 55 km/h (35 mph) if they are African buffalo.

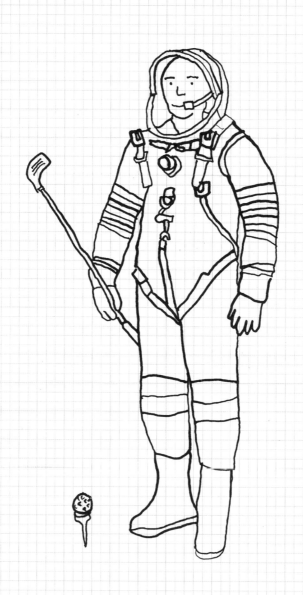

people power

Both men and women like the challenge of doing pointless things: running, jumping, hurling, chasing pieces of pigskin and other spherical objects over grass, and sometimes (especially in golf) emitting loud oaths at regular intervals. In most of these games, speed is important in some way.

Some psychologists argue that teams of players, ranging from nine in baseball to eleven in football, cricket and hockey, thirteen or fifteen in Rugby codes and eighteen in Australian rules football, represent the sizes of hunting groups in pre-agricultural human society, and that this explains the often anti-social behaviour of young male sports players when they are away from home.

Such theories do not really account for female team games, or the more individualistic activities like tennis, golf or marathon running, nor do they explain diving, darts, bowls or synchronised drowning. In all probability, there is more than one explanation. If you are the sort of person who likes football because it allows alpha males to belt each other, the notion of team-as-gang may appeal, and the idea of enclosing them in an arena may seem quite attractive.

Then again, as a former hockey player in a family of hockey players, I can certainly see elements of the hunt of small, roastable animals in the single-minded stick-waving pursuit of a small ball over turf.

Whatever the reason, most humans seem to enjoy competitive activities where speed is a key element, even if that speed is sometimes lethal.

the long-distance humans

Humans and pre-humans have been walking upright for about 4.5 million years. They have probably been running for most of that time, if only to get back to the shelter of trees around the African savanna. Even so, the quickest speed a human can reach is about 10.2 m/s (33.5 ft/s) for around 15 seconds, while horses, dogs, cattle, deer and antelopes can manage 15 to 20 m/s (50 to 66 ft/s) for much longer periods.

Humans do shine when it comes to endurance over long distances, and evolutionary biologists now think that running may have been a key part of our evolution. We are the only primate that engages in endurance running, and distance running is rare in animals other than social carnivores (like dogs and hyenas) and migratory ungulates (like wildebeest and horses).

In late 2004, biologist Dennis Bramble and anthropologist Daniel Lieberman argued in the heavyweight journal *Nature* that our anatomy looks the way it does because our ancestors were more likely to survive if they could run and keep on running. This was not necessarily to escape predators though: they may have needed to hunt animals or scavenge carcasses on the vast African savannas.

If they are right, running made us human – at least in an anatomical sense. *Australopithecus* was walking upright on two legs about 4.5 million years back, but fossil remains suggest that these early pre-humans could also travel through the trees. Then, about 1.5 million years ago, our genus, *Homo*, appeared on the scene with a very different body form.

Some new factor had to have emerged to give *Homo* short legs, long forearms, high permanently 'shrugged' shoulders, ankles that were less apparent, and more muscles connecting the shoulders to the head and neck.

If walking had not changed *Australopithecus* much over a whole 3 million years, then walking alone cannot explain why *Homo habilis*,

Homo erectus and, finally, our species, *Homo sapiens*, were so different. The nature of the changes, say scientists, favours the idea that something in the environment where the genus *Homo* lived made it important to be able to run for long periods of time.

The new humans combined reasonable speed on the ground with exceptional endurance, perhaps to chase down prey, or to be first on the scene to scavenge a carcass, a better prize in those days than a medal or a ribbon!

the marathon

The marathon race as we know it today originated with a battle at Marathon in 490 BC between Persia and Greece. Legend has it that an Athenian herald ran the 240 km (150 miles) to Sparta to seek their help against the Persian invaders. After the battle was won, the herald ran to Athens, just over 40 km (25 miles) away to announce the victory, and then dropped dead. As there were probably Persians on the main road, chances are he took a more difficult but shorter road, but who cares about facts?

The herald's name has not come down to us in a reliable form, and is given as either Thersius, Eucles, Philippides or Pheidippides. The last, Pheidippides, was the name given in surviving copies of a work written by Herodotus, who was born soon after the battle and wrote his account some 40 years later, but there is no mention in his Book VI of the run from Marathon to Athens. Pheidippides was also the name used in Robert Browning's 1876 poem about that first marathon run.

The race distance was 40.23 km (25 miles) for the first few races, but in London in 1908, the then Princess of Wales wanted her children to get a good view of the start, which was to be near Windsor Castle, so the start was set a further mile back from the finishing point. Then Queen Alexandra, the wife of Edward VII, wanted a good view of the finish, so it was moved further on, another 352 m (385 yd) to a point below the royal box. This distance, set by the mindless whims of selfish royalty, has remained fixed ever since at 42.186 km (26 miles and 385 yd).

At the time of writing, the marathon records for men and women were, respectively:

Haile Gebrselassie (Ethiopia)	Berlin, 2007, 2:04:26, 20.36 km/h (12.71 mph)
Paula Radcliffe (Great Britain)	London, 2003, 2:15:25, 18.75 km/h (11.7 mph)

from running footman to ultramarathoner

Robert Chambers was a publisher and an author. Because his house also printed a variety of religious works, he left his name off the title page when he published his *Vestiges of Creation*, which proposed a form of evolution, and his anonymity lasted for many years.

In his *Book of Days*, Chambers collected a miscellany of reminiscences, thoughts and ideas: judging by that work, he would be a popular blogger today. Published in 1869, it touches on many things, including the story of the running footman, who he said could cover 7 miles (about 11 km) in an hour or 60 miles (about 96.5 km) in a day; though he added that the roads were so bad that 5 miles (8 km) an hour was more normal. Footmen ran beside the coach, levering it out of potholes, and they also ran messages.

> . . . the Earl of Home, residing at Hume Castle in Berwickshire, had occasion to send his foot-man to Edinburgh one evening

on important business. Descending to the hall in the morning, he found the man asleep on a bench, and, thinking he had neglected his duty, prepared to chastise him, but found, to his surprise, that the man had been to Edinburgh (thirty-five miles) and back, with his business sped, since the past evening.

On another occasion, said Chambers, a footman ran 148 miles (238 km) in less than 42 hours to fetch medicine, including an overnight sleep before his return. In the end, better roads made the running footmen redundant. Still, their spirit lives on: here are some ultramarathon records (Castaneda's performance is unconfirmed).

100 km, Victor Ginko	1965, 8:38:7, 11.58 km/h (7.23 mph)
200 km, Zbigniew Klapa	1952, 19:55:07, 10.04 km/h (6.27 mph)
24-hour run, Jesse Castaneda	1976, 228.930 km, 9.54 km/h (5.96 mph)
24-hour run, Paul Forthome	1984, 226.432 km, 9.44 km/h (5.89 mph)

comparing males and females

My ultramarathon figures are all for men: why? Sir Roger Bannister broke the 4-minute mile in 1954, and the male record is now about 3 minutes and 45 seconds. All the same, if the male record holder and the female record holder were to run the mile at their best speeds, the woman would finish almost half a lap behind.

While a woman athlete may outrun a random man, average performance and elite performance favours the male of the human species. Men have a greater aerobic capacity (the rate at which the body can use oxygen to produce energy), they have more lean body mass, less fat and they have a greater cardiac output (the rate at which the heart pumps blood). They also have larger hearts, and their blood has more haemoglobin.

While it is common to separate horses by gender, fillies have won the Kentucky Derby and the only horse to win three Melbourne Cups was a mare, Makybe Diva. Mares and fillies may be seen in the winners' circle of other races at other times, even though they have few opportunities to outclass stallions and geldings.

Greyhounds aren't separated by gender, yet dogs, like horses and humans, show sexual dimorphism, with males usually larger and more heavily muscled than females. Male and female dogs, it seems, run equally well.

Nobody knows why humans are different, but scientists speculate that dogs evolved as a predator species and horses as a prey species, where males and females all needed to run. Humans have been tool users with gender-specific roles for maybe a million years, perhaps long enough for female running speed to be traded off in some way for other values. All the same, we need to take gender into account when we look at human performances.

the speed of a sprinter

Sprinting over a fixed course is by no means new. The earliest recorded race was the 180 m (590 ft) sprint of the 776 BC Olympics, which was won by a cook named Coroebus of Elis, but there were almost certainly foot-races long before that.

We give more acclaim to sprinters today than we do to any other athletes – which is possibly why we have seen so much drug abuse among elite sprinters in the last couple of decades. In Jesse Owens' day though, it was a bit different.

In one afternoon in 1935, Owens equalled one world record (the 100 yard (91.4 m) sprint), and set three new world records (long jump, 220 yard (201.2 m) sprint and 220 yard low hurdles), all in the space of just 45 minutes. In 1936, he took four gold medals at the Berlin Olympics, and though he was never snubbed by Hitler, he was most certainly never acknowledged by any US president before Gerald Ford.

In those days of amateur ideals, Owens had to make money any way he could after the Games were over, and one of these was to race against thoroughbred horses over short distances, relying on the horse being 'spooked' by the starter's gun. Even then though, he could be caught by the 100 yard (91.4 m) mark, so he raced over shorter distances, and never against a quarter horse.

A top sprinter today can run the 100 m (109.4 yards) in just under 10 seconds, a speed of 36 km/h (22 mph). Of course, that is from a standing (or crouching) start, so it is worth noting that in the fastest 15 m (16.4 yards), the top sprinters run at 45 km/h (27.9 mph), briefly. By comparison, a good middle distance runner will turn in an average speed of around 24 km/h (15 mph).

human jumpers

Leaving out hurdling and diving, there are four main forms of competitive jumping human beings undertake: the high jump, the long (or broad) jump, the hop-skip-and-jump (which gained dignity by becoming the triple jump), and the pole vault.

The high jump involves raising your centre of gravity as high as possible, while at the same time, getting all of your body parts over a bar that hangs precariously on two uprights. Success comes from running as hard as you can at the jump area, and then converting your forward energy into a vertical leap.

Your centre of gravity is a little over halfway up your height, somewhere not far from your navel. If you are 2 m (6.5 ft) tall for example, and clear the bar at that point, you have succeeded in raising your centre of gravity by about 90 cm (35.5 inches). Now by a neat bit of physics that we will come to in 'Leaping from a tall building', page 160, your upward speed must be the same as an object falling from 90 cm (35.5 inches). Your upward speed in this situation as you took off was 4.2 m/s (14 ft/s), just over 15 km/h (around 9 mph), less than half of the maximum speed a sprinter can achieve. The current world record of 2.45 m (2.68 yd) is equal to about 5.2 m/s (17 ft/s), almost 19 km/h (12 mph), of vertical speed at take off.

A top pole vaulter can clear 6 m (about 20 ft) which is equivalent to almost 10 m/s, or 36 km/h (22 mph), close to the world record for sprinters. The catch is that part of the energy comes from the run-up, but there is more energy still as the runner pushes forward when the pole digs in, and more energy from the jumper gaining leverage on the pole.

For male long jumpers, the approximate velocity (v) needed for a jump of (s) metres is (s + 2.23)/0.95, while for females, it is (s + 2.81)/0.99. Jesse Owens' 1935 record of 8.13 m (8.89 yd) implies an improbable speed of 11 m/s (12.03 yd/s), so use this guide with care!

running, leaping and hurling on the moon

Most people know that gravity on the Moon is about one-sixth of that on Earth, so they assume that Javier Sotomayor, who cleared the high jump bar at 2.45 m (8 ft) would be able to clear 14.7 m (48 ft) on the Moon – assuming he was in an enclosure and did not need a space suit of course.

If we leave out air friction, which would have to be there if the humans were breathing, athletes who throw things would probably get close to doing 6 times better, but there is a catch: running fast needs gravity, and this would affect some of the throwing events.

There is no easy way to do the analysis, but a sprinter who takes off out of the blocks needs to get a grip on the ground; under low gravity, the runner would probably lose contact with the ground, and be unable to keep those legs pushing forward. On top of that, some part of the movement of the arms and legs relies on a pendulum effect, and the period, the swinging time, of any pendulum depends on the gravitational forces acting on it.

In other words, there are good reasons to expect that running on the Moon would be harder than it is on Earth. Let us set that aside for a moment. Sotomayor's 2.45 m (8 ft) implies a vertical speed of 5.2 m/s (5.5 yd/s). If we take the acceleration due to gravity on the lunar surface as 1.62 ms^{-2}, we find that Sotomayor would raise his centre of gravity 8.35 m (27 ft), which would see him just clear around 9.5 m (31 ft), more than 5 m (16.5 ft) short of the predicted height.

While that would be fairly spectacular, we have to keep in mind that we made the dubious assumption that Sotomayor could run as fast on the Moon as on Earth. Don't bet on the high jump at any Lunar Olympics!

swimming records

Swimming in a continuous line is always more challenging than swimming laps in a pool. Just watch the swimmers as they tumble-turn and kick off, and you will realise that they gain quite a deal on each turn. This is why short-course (25 m or 27 yd) pools produce better results than long-course (50 m or 56 yd) pools.

There is probably an optimum-length pool, where a 'swimmer' could dive in, glide to the other end, turn, push off and glide back, over and over again, threading a distance much faster than even today's records. The difference between long- and short-course pools, however, is slight: Grant Hackett's 1500 m (640 yd) records for long- and short-course are 14:34.56 and 14:10.10, equivalent to 6.17 and 6.34 km/h (3.83 and 3.94 mph) respectively.

Alexander Popov swam the 50 m long-course in 2000 in 21.84 seconds, equivalent to 8.32 km/h (5.17 mph), while Roland Schoeman's 2006 record for the short-course 50 m was 20.98 seconds, an average speed of 8.58 km/h (5.33 mph).

Over a short distance, either of the sprinters could catch a jackass penguin. It is likely though that the jackass penguin could keep going at 7 km/h (4.4 mph), and if it could be trained to swim laps, that it would beat Grant Hackett over the 1500 m.

Women's speeds are slightly slower because their bodies are less streamlined. Against that, they have more fat under the skin (subcutaneous fat) which, although it raises them higher in the water, causes a greater loss of speed due to the lack of streamlining.

At the time this was written, Libby Trickett held the women's 50 m freestyle record with 23.97 seconds while Kate Ziegler holds the women's 1500 m freestyle long-course record with 15:42.54. These speeds are equivalent to 7.51 and 5.73 km/h (4.66 and 6.56 mph).

all downhill

Ever since Jack and Jill – and probably for some time before that – humans have known that one way to go very fast is to roll downhill. In some societies, drunken young men are said to have placed old women in barrels and rolled them down hills, and if low-grade louts can work out that things go fast downhill, we should not be surprised that sports enthusiasts worked it out.

Fortunately for ice skaters, there isn't too much risk of sliding downhill as there are very few sloping bodies of ice that are smooth enough to skate on, but this has not stopped people sliding downhill on ice. It is fortunate that luge and bobsled riders are not entirely devoid of common sense, which is why their courses have curly bits to slow them down (or tip them over), and soft bits at the end.

The fastest recorded luge rider was Norwegian Asle Strand who reached 137.4 km/h (85.4 mph) at Tandådalens Linbana, Sälen, Sweden in May 1982. This was based on photo-timing. Street luges are claimed to have reached speeds of 120 km/h (75 mph), and there are dubious claims of 160 km/h (100 mph) rocket-assisted street luges that probably owe something to a certain JATO-powered Chevy Impala legend that we will visit later ('The JATO Chevy Impala urban myth', page 90).

Downhill skiing goes well past that sort of speed, and as an accepted sport, has the advantage of officials who can confirm performances. Philippe Billy is the world record holder, having hit 243.902 km/h (151.586 mph) at Vars in France in 1997.

Until they can find a way to skate downhill, speed skaters will be limited: at the moment, the record for 500 m is held by Joji Kato, who covered the distance in 34.3 seconds, a speed of 14.58 m/s (16 yd/s), or 52.48 km/h (32.6 mph).

racquet and court games

Squash players will tell you that being hit in the eye by a squash ball is far more dangerous than being struck by a tennis ball. The reason, they say, is that a squash ball is close in size to the eye socket, and the hydraulic pressure placed upon the eye is diabolical. No other small ball wreaks as much havoc, but there are two other factors that come into play: speed and distance.

Watch a top-level game of tennis on television, and somewhere in the background, you'll see a digital display from a Doppler radar to give you the speed of a serve. In tennis, the fastest ball is the serve, simply because the player has time to line up on the ball and accelerate the racquet through a large arc.

Andy Roddick has been reliably clocked serving at 240 km/h (150 mph), though using less reliable methods, 'Big Bill' Tilden was credited with 263 km/h (163.3 mph) in 1931. Other top players in the past were credited with a more believable speed of around 180 km/h (110 mph).

Much like tennis, the Basque game of jai alai uses a court, but the ball is thrown from a curved implement, rather than hit. The ball travels amazingly fast, at 84 m/s (92 yd/s), 302.6 km/h (188 mph). Charles Lindbergh flew the Atlantic slower than that, and no land vehicle reached that speed until 1927!

There is a great deal of difference between ping pong and table tennis, just as there is between hit-and-giggle and competitive tennis, or the dog-paddle and swimming the English Channel. A good table tennis player like Lark Brandt can reach 112 km/h (70 mph), but the ball is too light to inflict real damage, and if Fu Haifeng can hit a badminton shuttlecock at 320 km/h (200 mph), it slows down too fast to do real harm. Squash balls remain the big threat.

A squash ball weighs 24 grams (plus or minus 1 gram), just under an ounce, and is often hit from very close range. A tennis ball weighs more than double that, but it is larger and has more air resistance as it travels much of the length of the court, with air friction slowing it, before it reaches the opposing player who can see it coming. The squash ball also travels much faster: in 2004, John White was clocked serving at

Figure 5. Projectile speeds

Projectile details	Mass (grains)	Mass (kg)	Velocity (ft/s)	Velocity (m/s)	Energy (foot pound)	Energy (J)
.22 conical ball cap	29	0.0019	350	107	8	11
Squash ball at 150 km/h (90 mph)	370	0.0240	132	40	14	19
.22 bulleted breech cap	18	0.0012	700	213	20	27
.22 conical ball cap	20	0.0013	700	213	22	30
Squash ball at 270 km/h (170 mph)	370	0.0240	249	76	51	69
.22 short	29	0.0019	1045	319	70	95
.22 long	29	0.0019	1240	378	99	134

270 km/h (170 mph), and many elite players, both male and female, get close to that mark. The official standard for testing squash goggles uses a ball travelling at 150 km/h (around 90 mph), and from a range of 1 to 2 m (3 to 6 ft)! That speed would be reached regularly by your average good player.

Max Moorhouse of i-Mask, a manufacturer of protective squash goggles, tells me that the World Squash Federation warns that the impact of a squash ball at this speed is equivalent to that of a .22 calibre bullet.

I doubted this, but Figure 5 bears this approximation out for all but top-end ammunition. Watch out!

death and mayhem on the diamond and the field

Deaths in baseball and cricket from direct hits are mercifully rare. Legend has it that Frederick, the oldest son of King George II, died after being hit on the head by a cricket ball, but he actually died of an abscess on the lung. Batter Ray Chapman *was* killed in 1920 when hit by a 'beanball' from the Yankees' Carl Mays. Although not all life-threatening, a number of baseball players in the USA sustain injuries from errant balls each year.

Batters are required to be protected by a helmet, but the same does not apply to pitchers, and each year, around 20 US college pitchers and infielders are hit and removed from the game after balls strike them. By the time they get to college, the players have largely reached their full adult strength, but unfortunately some still lack the sporting skills that come with maturity.

A cricket field is larger than a baseball field, and the ball may be driven to any part of it. In baseball, the target area for scoring purposes is far smaller, making it hard to hit the ball far enough to allow for a clear run between two consecutive bases, let alone further. That makes it imperative for the batter to hit the ball hard and fast, preferably over the outfield fence for an automatic home run, but failing that, into a gap on the field.

Even if the pitch comes in at less than 160 km/h (100 mph), the ball can typically 'exit' the bat at more than that speed. There is no exact science that can explain the actual speed, but factors include the speed of the bat, the weight of the bat, and how much time the bat has to transfer energy to the ball, something which depends on the material of the bat and the material of the ball.

One of the available gap areas is close to the pitcher, who rarely has time to react and catch it. The ball may well be snapped up by centrefield, but remember that college players are young and inexperienced.

They have a limited time to react, and as young men, the batters will be putting all of their strength into truly smiting the ball. On occasions, it may sail dangerously close to, or straight at, the pitcher. Infielders are also at risk, but they are further away, and have more time to react to a threat.

During the past decade or so, a great deal of research has been undertaken in the United States into the factors that make for a 'fast' ball, so the rest of this discussion will be in 'imperial' units. Bats made of graphite or metal usually deliver a faster ball, all other things being held constant. And while bats can also differ in size, a longer bat actually produces a slower ball than a shorter one.

Under the rules, the length in inches and the weight in ounces are linked: the difference cannot be greater than 3, so a 33 inch (84 cm) bat must weigh at least 30 ounces (850 grams). Laser studies reveal that a 30 ounce bat reaches a top speed of 63 mph (101.4 km/h), while a 28 ounce (795 gram) bat swings at 65 mph (105 km/h).

There is a down side to this, though. At Mississippi State University, where some of the research has taken place, Dudy Noble Field has a 20 ft (6 m) centrefield fence, 390 ft (119 m) from the home plate. If a batsman wants to clear this, the ball needs to leave the bat at 105 mph (169 km/h). If restrictions were put into place to limit balls to 93 mph (150 km/h), researchers expect that the ball would fall 20 ft (6 m) short. Making the game safer for pitchers and infielders may take away one of its best features, although more recent proposals have raised the suggested limit somewhat.

One curious side issue: popular wisdom says it is easier to hit a fastball further than a curve ball, but it turns out the opposite is true. A curve ball comes in with a topspin which reverses when it is hit, providing a backspin which carries it further. A fastball has a backspin which becomes a topspin when it is hit, so the ball will not carry as far.

the physics of home runs

The popularity of baseball as a college sport in the United States guarantees that at least some members of a university physics department are likely to be interested in baseball. A few may verge on having what a dispassionate outsider would call an obsessive concern with the sport. This happens less often with cricket.

In cricket, hitting the ball over the boundary is six runs. It's nice when that happens, but it is possible to score runs in other ways – a match of any length will involve at least some hundreds of runs. In baseball, if the bases are loaded, a home run brings all of the players home and scores four runs in a game where the winning margin is often less than that.

The result is that the home run is pursued with fervour, and many otherwise rational men and women will devote large parts of their lives to working out what makes the ball fly the furthest when it comes off the bat. For those of us who do not care as much for baseball, the basic principles are still relevant when it comes to considering other games where a ball is struck.

The forces applied are extreme, because only a monumental blow will send the ball off with enough speed to clear the fence. The ball may be compressed up to half its diameter before it springs back again, and bats can shatter under an impact that only lasts perhaps half of one thousandth of a second.

The distance travelled by a ball hit on the perfect angle will approximate to v^2/g, where g is the acceleration due to gravity – if we ignore wind resistance and spin on the ball. In point of fact, each of these is important, but from that deceptively simple formula, it follows that the velocity of the ball is vital.

We can think of the ball as a simple spring. When it is hit, the ball compresses, and as it springs back, the potential energy stored as compression is converted into kinetic energy. Some energy is converted to heat: this is hard to detect in a baseball, but in a squash ball, where the ball is hit repeatedly over a short period of time, the behaviour of the squash ball changes in play, as the ball warms up.

Some energy is lost to the ball when the bat recoils, so a heavier bat, with less recoil, is desirable. As we have seen, though, a heavy bat does not swing as fast, and a compromise is needed.

Like a tennis or squash racquet, a bat has a sweet spot, and when the ball is struck there, the bat vibrates less, and once again, more energy is transferred to the ball. Aluminium bats are stiffer than wood bats, and that gives them a larger sweet spot. This makes aluminium bats more forgiving, which is part of the reason why they are not allowed in the major leagues.

Terry Bahill, a professor of systems and industrial engineering in Arizona, argues that outlawing aluminium bats in college baseball would produce faster-batted ball speeds, endangering pitchers. He says that as a rule, lighter bats are better for hitters, but if a batter wants the same length in a wooden bat that is available in aluminium, the bat will be heavier, which means that a struck ball will travel a little faster from the wooden bat.

In the last 30 years of the twentieth century, the actual baseballs themselves got livelier as well. By examining a number of donated baseballs, researchers inferred that recent balls use synthetic materials instead of wool in the windings, and more rubber in the core.

The American manufacturer's standard requires that balls be fired from an air cannon at 93 km/h (58 mph) against an immobile northern white ash (baseball bat timber) surface. The resiliency is measured from the rebound, and must lie in the range from 51% to 58%. By way of comparison, putty has a resiliency of zero, and a 'Superball' has a resiliency of about 90%.

Baseball aerodynamics were first studied experimentally in the 1950s by helicopter pioneer Igor Sikorsky, but his results were lost for over

40 years before being found in a Cape Cod attic. Sikorsky's data dealt with baseball pitches over 153 km/h (95 mph), while other experiments looked at pitches up to 64 km/h (40 mph). No data covered intermediate spin rates, typical of college and semi-professional baseball these days.

Spinning a ball always makes it curve, because the ball's surface is moving faster on one side of the ball than on the other. Recent tests have also shown that scuffing of the ball plays a role, as expected, and that whether it is a two-seam or four-seam ball has an effect also.

Looking at it another way, when an aeroplane wing travels through the air, the upper surface is curved, and air passing over it needs to travel a greater distance than air passing under the wing. This produces lift that pushes the plane upwards. With a spinning ball, the side of the ball that is moving forwards has a greater relative velocity to the air it passes through, creating lift that pulls the ball towards that side.

The forces are due to a few factors: spin, 'roughness' of the ball, and whether it is scuffed, causing turbulence. A four-seam ball curves up to three times as much as a two-seam ball with the same velocity. This is not surprising because the extra seams confer more roughness.

fast balls on grass

The comparison between cricket and baseball is an interesting one as they are in the same game 'family'. Most observers concentrate on the differences, but there are also some interesting similarities. The distance from crease to crease on a cricket pitch is 17.7 m (58 ft), while the distance from mound to plate in baseball is 18.4 m (60.5 ft).

The throwing action of a pitcher in baseball and a bowler in cricket are quite different, yet the two balls travel at remarkably similar speeds. According to radar measurements, Mark Wohlers threw the fastest-ever pitch, with a speed of 166 km/h (103 mph). Against this, the fastest bowler in modern cricket was Pakistan's Shoaib Akhtar, who put up a speed of 161.3 km/h (100.2 mph) when he was bowling against England in the 2003 Cricket World Cup.

There are differences as well: it is normal for the slightly lighter, smaller cricket ball to strike the pitch once and bounce up – this slows the cricket ball down somewhat, but does make it fly off unexpectedly sometimes. There may, however, have been faster balls in the past, going on the evidence of E. V. Lucas, who wrote in 1904 in *Highways and Byways in Sussex*:

> At the neighbouring village of Stoughton, whither I meant to walk (since an inn is there) was born, in 1783, the terrible George Brown – Brown of Brighton – the fast bowler, whose arm was as thick as an ordinary man's thigh. He had two long stops, one of whom padded his chest with straw. A long stop once held his coat before one of Brown's balls, but the ball went through it and killed a dog on the other side. Brown could throw a 4-½ oz. ball 137 yards, and he was the father of seventeen children. He died at Sompting in 1857.

Now just try and show me a baseballer who can match that!

the flight of the golf ball

In 1971, the astronaut Alan Shepard played the first-ever golf stroke in space when he struck a ball on the Moon. He played with an improvised club – a Wilson 6-iron head attached to a lunar sample scoop handle, and later claimed that the second of two balls went 'miles and miles'.

In 2006, cosmonaut Mikhail Tyurin played a shot of sorts from the International Space Station. He did so one-handed, something that would hardly be approved of on most courses on this planet. Worse still, US astronaut Michael Lopez-Alegria held onto Tyurin's feet while he made the stroke. OK, so it stopped him from accidentally flying off into space, but it would never be accepted at the Royal and Ancient!

The ball weighed far less than a standard ball, just 3 grams (0.105 oz), and Tyurin hit it with a gold-plated 6-iron as part of a stunt to promote a Canadian golfing equipment company. The shot was shanked, but the ball was still expected to burn up in the atmosphere after travelling for several days. With that sort of time frame, even a gentle tap would result in what the company claimed as 'the longest drive'.

There are many who claim this title, though quite a few of the alleged records were achieved far from the golf course: strokes have been played along airstrips, from hills overlooking frozen rivers, and in other conditions that would artificially enhance the 'carry' of the ball.

The flight of a ball on the Moon, or from a space station, would differ remarkably from one hit on a golf course on Earth. The stroke itself uses muscle power to accelerate the club head towards the ball, but the force of gravity also plays a small part. The lower gravity on the Moon and the 'lack of gravity' in space (there *is* gravity in space, you just don't notice it) play a bigger part.

Then there is the influence of the air. From the space station, the ball would have followed a complex curve, shaped first by the Earth's gravity and then by the drag from the wisps of atmosphere that it ran into. On the Moon, the ball would describe a parabola, moving forward as fast when it touches down as it moved when it left the tee.

On Earth, things are different. The ball climbs steadily, and sometimes even begins to climb at a steeper angle before it slows and then drops, more or less straight down to the fairway.

This odd path is predictable from the way the ball starts out. With a perfect on-centre hit, the ball climbs at a shallow angle, with a backspin of as much as 3000 rpm, travelling at around 250 km/h (160 mph). By the time it falls to the grass, the ball's forward motion will be less than a third of that. Not all has been lost though because some of the energy carried by the ball was used to keep it aloft so it could move further forward.

The secret is an aerodynamic effect called Magnus lift, and this is why golf balls are now dimpled. They help form a turbulent layer underneath the ball that keeps it floating. This was not something physicists thought up: it all began when golfers noticed that their older battered balls went further. Speed was a key factor, but a dimpled ball was better!

Still, for most aspects of golf, speed (or lack of) is all-important. When it comes to the green for example, there is a severe speed limit for a putt. If your ball is travelling faster than 4.6 km/h (2.9 mph), and it hits the hole squarely, it will cross the cup before it drops far enough to be stopped by the opposite lip. If you want your ball to go in, rather than hopping up and over, that is one speed you must not exceed.

fast motion

The fastest goal in football (soccer) may be as short as 3.17 seconds, but in serious football, Roy Maakay's 10-second goal for Bayern Munich in a Champions League match against Real Madrid has to rate highly. In women's football, the fastest goal in World Cup competition was scored in 1991 against Japan in 30 seconds by a Swede, Lena Videlull.

Golf is normally a game played for relaxation. In 1981, Steve Scott, then the US 1500 m (1640 yd) record holder, completed 18 holes on a regulation course in 29 minutes, 33.05 seconds at Miller Golf Course in Anaheim, California. He posted a creditable (under the conditions) 92.

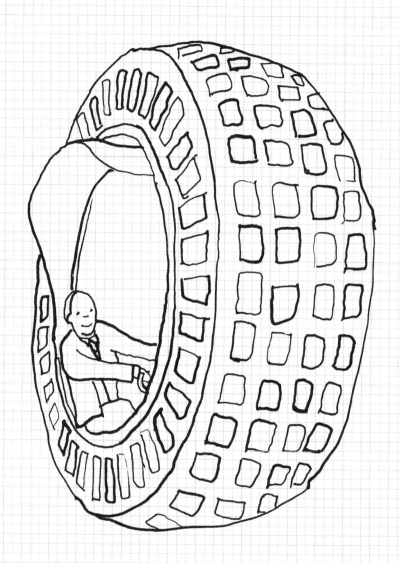

fast track

Something happened around 1825, because people began quoting the speed of things. Perhaps time-pieces became more accurate, or maybe it became easier to measure distances, allowing more precise speeds to be calculated. It probably had to do with moving faster.

The notion of speed was old, but measuring it as a rate wasn't. In the late 1700s, the speeds of carriages were measured in 'miles an hour'.

In July 1802, the *New York Herald* reported an Englishman named Shaw who travelled 172 miles (about 277 km) in 12 hours, 'at the astonishing rate of three minutes forty-one seconds per mile, being upwards of 16¼ miles per hour'.

The term 'miles an hour' then slowly disappeared, though it had been used since 1687. It continued to appear occasionally, mostly about carriages and coaches. In 1816, *The Times* reported a Montgolfier balloon moving over London at '50 miles per hour'. There are two uses of the phrase in 1819, two in 1823, and many more for every year from 1825 onwards.

With trains running on different tracks, and ships crossing the Atlantic at different times, there could be no races, so the idea of reducing rates to a common standard made more sense. It took time for records to start being recorded in terms of speed, because the time elapsed to sail from Liverpool to Sydney, or Plymouth to Boston was easier to take in. Sailors preferred a speed in knots, but other speeds were in miles per hour (and later, kilometres per hour).

By the 1890s, even human speeds began to be quoted in miles per hour, and nowadays, when we hear that a good hamster sprint covers 9 metres (30 ft) in 38 seconds, we reach for our calculators to reveal that this is equal to 0.86 km/h (0.54 mph).

the ride of the peloton, the art of the breakaway

There are many ways to become a cycling tragic. Where I live, the daylight Tour de France is happening during our night, so I can watch hours of rural France wheeling by, with views offered from the front of the race, the back of the race, in the middle of the pack, known as the peloton, and from the helicopters flying overhead. I began watching for the scenery and got hooked on the cycling.

Coming to watch world-class cycle racing as a raw beginner leaves a few puzzles in the mind, and it is sometimes hard to realise how much intelligence is needed. For much of each stage, most riders seem to just amble along, with time to chat to the riders around them. The key term here, though, is world-class.

While there have been a few world-class drug cheats in the past few years, most of the Tour de France riders are drug-free athletes. The peloton travels at a savage 45 to 50 km/h (28 to 31 mph). But if they are such athletes, ambling along seems a bit foolish, and the new watcher must wonder why some of them don't make a real effort, especially at the end.

There are several part-answers: first, by staying in a bunch, all riders except the very front one or two in the peloton gain a slipstreaming advantage. If you watch closely, you will see the peloton's stronger riders move up to the front, and take the lead for sometimes just a few hundred metres (yards), before slipping to the side and dropping back down through the pack again.

The second is that in major races, if you cross the line within one second of the rider in front of you, you get the same time as the front rider. So long as the peloton is bunched together, the last rider in the crowd gets the same time as the first rider, even though that rider may have crossed 20 or more seconds earlier. It is more important to stay with the group than to really push along, or ahead.

At the front of the pack though, things are a little different because sometimes a small group of riders will try to break away, and if there are enough of them, they may be able to form an independent group that can stay ahead of the peloton. The escape group has to work harder, and often one or more will drop away and allow the peloton to catch up.

At the end of the race, teams come into their own, leading out their specialist sprinters, with one rider after another taking the lead and hauling the group along at high speeds, usually in a town, sometimes with tram rails, cobbles or other distractions. At this point, the riders, other than the sprinter, are riding above their aerobic thresholds, but in the last 200 m (656 ft), the specialist sprinter takes off, hitting close to 70 km/h (43 mph).

Over longer distances, speeds drop. In an individual time trial, generally over flat ground with no drafting allowed, the record speed in the Tour de France was set in 2005 by Dave Zabriskie when he rode a 19 km (11.8 miles) stage at 55 km/h (almost 35 mph). Team time trials can involve drafting and can be faster.

When you look at a whole Tour de France, the mountain climbs slow riders down (even if they often hurtle down the far sides at speeds of up to 75 km/h (47 mph). Lance Armstrong's 2004 winning performance involved completing the race in 83 hours, 36 minutes and 2 seconds in the saddle, giving an official average of 41 km/h (25 mph). Two more riders came in, a little over 6 minutes later, a gap of 4 km (2.5 miles) after riding some 3500 km (2174.8 miles) in total.

speed and leisure

William Henry Davies (1870–1940), paid a price for his pursuit of speed. Born in Wales, he visited North America as a young man, and made his way around by 'jumping rattlers'. While he was in Canada, he lost a leg in an accident and wore a wooden leg thereafter. He returned to Britain in 1905, and gained recognition as a poet. Here, he poses a question for all devotees of the fast lane.

Leisure

What is this life if, full of care,
We have no time to stand and stare.
No time to stand beneath the boughs
And stare as long as sheep or cows.
No time to see, when woods we pass,
Where squirrels hide their nuts in grass.
No time to see, in broad daylight,
Streams full of stars, like skies at night.
No time to turn at Beauty's glance,
And watch her feet, how they can dance.
No time to wait till her mouth can
Enrich that smile her eyes began.
A poor life this if, full of care,
We have no time to stand and stare.

the hunt for the world's fastest motor vehicle

The first attempt at setting a motor vehicle record took place in France in December, 1898. Gaston de Chasseloup-Laubat was clocked over a flying kilometre (where the car crosses the start, already at full speed) at 57 seconds in a car called a Jeantaud. This equates to 63.15 km/h (39.24 mph), but the idea of rates seems to have been less interesting than the elapsed time back then.

In January 1899, Camille Jenatzy, a Belgian, reached 66.66 km/h (41.42 mph) over a flying kilometre, but his triumph was short-lived, as Chasseloup-Laubat promptly countered with 70.31 km/h (43.69 mph). Ten days later, Jenatzy hit the lead with 80.35 km/h (49.93 mph).

In March 1899, Chasseloup-Laubat reached 92.78 km/h (57.65 mph), and in late April, Jenatzy drove his car at 105.88 km/h (65.79 mph). The two then disappear from the record books, but Jenatzy's record stood until April 1902, when Leon Serpollet reached 120.8 km/h (75.06 mph).

Amazingly, not one of these record-breaking vehicles used internal combustion: Chasseloup-Laubat and Jenatzy both used electric cars, while Serpollet's vehicle was steam driven. In the very early twentieth century, internal combustion engines simply could not deliver the power that was needed.

The world land-speed record was about to get a great deal noisier: three petrol-driven Mors vehicles took the record during 1902, but under modern rules, where a new record speed had to be more than 1% faster than the previous record, two of them would have been disallowed.

In early 1904, the record was 147.05 km/h (91.37 mph) and the magic 'ton', 100 mph (about 161 km/h), was cracked twice in 1904, with A. Macdonald holding the record with 168.42 km/h (104.65 mph) by early the following year. The end of 1905 saw the record reach 175.44 km/h (109.01 mph), while 1906 saw F. Marriott set a record of a massive 205.44 km/h (127.65 mph).

The record held until 1910 when B. Oldfield reached 211.98 km/h (131.72 mph). Then, in 1914, this was paradoxically eclipsed by a slower record because of the new rule which required two runs in opposite directions, within one hour, in order to eliminate effects of wind and slope. L. G. Hornsted drove two flying one-mile runs at 206.25 km/h (128.16 mph) one way and 193.57 km/h (120.28 mph) in the other, for an average of 199.70 km/h (124.10 mph).

There was a break during and after World War I, even in the United States, which was neutral until 1917, and so Hornsted's record stood until 1922, when K. L. Guinness surpassed both Oldfield's and Hornsted's records under the new rules.

There were 22 more records set before World War II began in 1939. The last was on the 23 August 1939, less than a fortnight before war broke out. J. Cobb's record of 595.04 km/h (369.74 mph), set at Bonneville, held until 1947, when Cobb raised the speed to 633.79 km/h (393.82 mph).

This stood until 1963, when Craig Breedlove broke 657.114 km/h (408.31 mph). The record changed ten times during the period from 1963 to 1965, but that year signalled the end of the era of the internal combustion engine. Late in the year, Breedlove hit 966.961 km/h (600.842 mph) in a turbojet car, and all future land-speed records would be set in such vehicles.

Figure 6 gives speeds in km/h *only* for the kilometre course, but some figures were also quoted for the mile course.

great land speed record crashes

Over the years, a number of land-speed record attempts have led to catastrophic disintegrations. In some cases, the drivers were killed or seriously injured, but the lucky ones were those who were able to walk away. Sir Malcolm Campbell was one of the few determined record hunters who survived, and his son, Donald, was lucky to survive a crash at 579.4 km/h (360 mph).

When he came to, Donald immediately asked about repairing the car, and manufacturer Sir Alfred Owen declared: 'If Campbell has the guts to carry on, I'll build him another car.' Donald Campbell was killed in 1967 while pursuing the water speed record.

In 1907, Fred Marriott was pursuing the land speed record on the beach at Daytona in the Stanley 'Rocket', a steam-powered racing car. At something like 300 km/h (190 mph), he hit a slight depression that he thought the car could handle. This was not the case, and as he said later, the car took off like a kite, stayed in the air for about 30 m (100 ft), and broke up on landing.

The boiler of the car, with an incredible steam pressure of 1300 psi (about 90 atmospheres), travelled a further 300 m (1000 ft) down the beach. The rear end of the car buried itself in the sand, the front end, with the driver, headed for the sea. He ended up with his head under water, several broken ribs, a cracked sternum, a sliced scalp, one eye forced out of its socket and resting on his cheek, and he was bruised all over. A doctor eased the eye back into its socket, he was stitched, strapped and patched – and he survived! A month later, he was back in a car, setting a new speed record for steam cars.

Other challengers for the record were not so lucky. J. G. Parry-Thomas held the land speed record twice, but he was decapitated by a broken chain at Pendine Sands in Wales in 1927.

Craig Breedlove may have preferred not to achieve one record he notched up in 1964, when he entered the *Guinness Book of Records* for the world's longest skid. At the end of a run, his parachute failed to deploy, and he skidded for 8 km (5 miles) before passing through a row of telephone poles and into a brine pond, still doing 300 km/h (190 mph). From the information provided in 'Stopping distances' (see page 92), we can estimate that the brine pond deprived Breedlove of about another 5 km (3 miles) of skid. Undaunted, Breedlove returned to beat the 600 mph (about 960 km/h) barrier.

No two crashes have the same cause. Richard Noble's Thrust 1 was destroyed in 1977 when a wheel-bearing failed as he roared up the runway of RAF Fairford in Gloucestershire at 320 km/h (200 mph).

Figure 6. Land speed records

Date	Place	Country
18 December 1898	Achères	France
17 January 1899	Achères	France
17 January 1899	Achères	France
27 January 1899	Achères	France
4 March 1899	Achères	France
29 April 1899	Achères	France
13 April 1902	Nice	France
5 November 1902	Ablis	France
5 November 1902	Dourdan	France
17 November 1902	Dourdan	France
17 July 1903	Ostend	Belgium
5 November 1903	Dourdan	France
12 January 1904	Lake St. Clair	USA
27 January 1904	Daytona Beach	USA
31 March 1904	Nice	France
31 March 1904	Nice	France
25 May 1904	Ostend	Belgium
21 July 1904	Ostend	Belgium
13 November 1904	Ostend	Belgium
24 January 1905	Daytona Beach	USA
30 December 1905	Arles	France

Driver	Vehicle	Type	km/h
Gaston de Chasseloup-Laubat	Jeantaud	Electric	63.15
Camille Jenatzy	Jenatzy	Electric	66.66
Gaston de Chasseloup-Laubat	Jeantaud	Electric	70.31
Camille Jenatzy	Jenatzy	Electric	80.34
Gaston de Chasseloup-Laubat	Jeantaud	Electric	92.70
Camille Jenatzy	La Jamais Contente	Electric	105.88
Leon Serpollet	Serpollet	Steam	120.80
William K. Vanderbilt	Mors	IC	122.44
Henri Fournier	Mors	IC	123.28
M. Augières	Mors	IC	124.13
Arthur Duray	Gobron-Brillie	IC	134.32
Arthur Duray	Gobron-Brillie	IC	136.36
Henry Ford	Ford (on ice surface)	IC	147.05
William K. Vanderbilt	Mercedes (not recognised in Europe)	IC	148.54
Arthur Duray	Gobron-Brillie	IC	142.85
Louis Rigolly	Gobron-Brillie	IC	152.53
Pierre de Caters	Mercedes	IC	156.50
Louis Rigolly	Gobron-Brillie	IC	166.66
Paul Baras	Darracq Gordon Bennett	IC	168.21
Arthur MacDonald	Napier	IC	168.42
Victor Hémery	Darracq V8 Special	IC	175.44

Figure 6. Land speed records (continued)

Date	Place	Country
26 January 1906	Ormond Beach	USA
24 January 1907	Ormond Beach	USA
6 November 1909	Brooklands	UK
23 March 1910	Daytona Beach	USA
24 June 1914	Brooklands	England
17 May 1922	Brooklands	England
6 July 1924	Arpajon	France
12 July 1924	Arpajon	France
25 September 1924	Pendine Sands	Wales
21 July 1925	Pendine Sands	Wales
16 March 1926	Southport	England
27 April 1926	Pendine	Wales
28 April 1926	Pendine	Wales
4 February 1927	Pendine Sands	Wales
29 March 1927	Daytona Beach	USA
19 February 1928	Daytona Beach	USA
22 April 1928	Daytona Beach	USA
11 March 1929	Daytona Beach	USA
5 February 1931	Daytona Beach	USA
24 February 1932	Daytona Beach	USA
22 February 1933	Daytona Beach	USA
7 March 1935	Daytona Beach	USA

Driver	Vehicle	Type	km/h
Fred Marriott	Stanley Rocket Racer	Steam	205.45
Glenn Curtiss	Curtiss V8 40 hp (30 kW) motorcycle	IC	219.31
Victor Hémery	Benz (fastest IC vehicle)	IC	202.68
Barney Oldfield	Benz	IC	211.94
L. G. Hornsted	Benz (first two-way record)	IC	199.70
Kenelm Lee Guinness	Sunbeam	IC	215.17
René Thomas	Delage La Torpille	IC	230.47
Ernest A. D. Eldridge	Fiat	IC	234.98
Malcolm Campbell	Sunbeam 350HP Blue Bird	IC	235.21
Malcolm Campbell	Sunbeam	IC	242.79
Henry Segrave	Sunbeam	IC	245.10
John Parry-Thomas	Higham	IC	272.46
John Parry-Thomas	Higham	IC	275.17
Malcolm Campbell	Bluebird II	IC	281.44
Henry Segrave	Sunbeam	IC	326.66
Malcolm Campbell	Bluebird III	IC	333.05
Ray Keech	White Triplex	IC	334.02
Henry Segrave	Irving-Napier Golden Arrow	IC	372.66
Malcolm Campbell	Campbell-Napier-Railton Bluebird	IC	396.03
Malcolm Campbell	Campbell-Napier-Railton Bluebird	IC	408.73
Malcolm Campbell	Campbell-Railton Bluebird	IC	438.48
Malcolm Campbell	Campbell-Railton Bluebird	IC	445.50

Figure 6. Land speed records (continued)

Date	Place	Country
3 September 1935	Bonneville Salt Flats	USA
19 November 1937	Bonneville Salt Flats	USA
27 August 1938	Bonneville Salt Flats	USA
15 September 1938	Bonneville Salt Flats	USA
16 September 1938	Bonneville Salt Flats	USA
23 August 1939	Bonneville	USA
16 September 1947	Bonneville	USA
5 August 1963	Bonneville Salt Flats	USA
17 July 1964	Lake Eyre	Australia
5 October 1964	Bonneville Salt Flats	USA
7 October 1964	Bonneville Salt Flats	USA
13 October 1964	Bonneville Salt Flats	USA
15 October 1964	Bonneville Salt Flats	USA
27 October 1964	Bonneville Salt Flats	USA
2 November 1965	Bonneville Salt Flats	USA
7 November 1965	Bonneville Salt Flats	USA
13 November 1965	Bonneville	USA
15 November 1965	Bonneville Salt Flats	USA
23 October 1970	Bonneville Salt Flats	USA
4 October 1983	Black Rock Desert	USA
15 October 1997	Black Rock Desert	USA

Driver	Vehicle	Type	km/h
Malcolm Campbell	Campbell-Railton Bluebird	IC	484.62
George Eyston	Thunderbolt	IC	502.11
George Eyston	Thunderbolt	IC	556.02
John Cobb	Railton Special	IC	563.59
George Eyston	Thunderbolt	IC	575.07
John Cobb	Railton	IC	594.97
J. Cobb	Railton	IC	634.4
Craig Breedlove	Spirit of America (3-wheeler)	Turbojet	657.11
Donald Campbell	Bluebird CN7	Gas-turbine	644.96
Tom Green	Wingfoot Express	Turbojet	668.03
Art Arfons	The Green Monster	Turbojet	699.03
Craig Breedlove	Spirit of America	Turbojet	754.33
Craig Breedlove	Spirit of America	Turbojet	846.86
Art Arfons	The Green Monster	Turbojet	875.70
Craig Breedlove	Spirit of America - Sonic 1	Turbojet	893.97
Art Arfons	The Green Monster	Turbojet	927.87
B. Summers	Goldenrod (piston-engined, wheel-driven)	IC	658.67
Craig Breedlove	Spirit of America - Sonic 1	Turbojet	966.96
Gary Gabelich	Blue Flame	Rocket	1014.52
Richard Noble	Thrust2	Turbojet	1019.47
Andy Green	ThrustSSC	Turbofan	1227.99

The car flipped over three times and was a write-off, but Noble survived to take Thrust 2 close to supersonic in 1983. Later analysis indicated that if that car had gone a mere 11.3 km/h (7 mph) faster, it would have become airborne, and almost certainly crashed.

In 2006, television presenter Richard Hammond was working on the BBC *Top Gear* program when he drove 'Vampire', the car Colin Fallows had used to establish the British land speed record of 483.3 km/h (300.3 mph) in 2000. On the second of three laps, Hammond reached 505.3 km/h (314 mph), but on the third lap, he crashed. Since record attempts require an average of two trips in opposite directions, Hammond missed the record, but he got to keep his life.

The lust for speed is such that some people will take almost any risks to go faster. In one or two cases, though, the facts do not bear out the stories that are told. The 'JATO-assisted Chevy Impala' (below) remains one such beautifully crafted yarn.

the jato chevy impala urban myth

This never happened, but it not only deserves to have happened, it sounds like it *might* have happened. That is probably why you still find versions of it circulating around the internet. If you need further evidence that it never happened, go to www.snopes.com, and search for 'JATO Chevy'.

Part of the attraction is the magnificent foolishness of the attempt, a real boys' toys yarn, but the clincher is probably the gruesome details of the evidence, even down to the teeth and bone fragments.

Once again: the following story is not true. It is an urban legend. What is true is that a JATO (Jet Assisted Take-Off) Unit is not a jet at all, but a rocket. During World War II, when the JATO units were being developed, the word 'rocket' carried a certain aura of danger about it, so the boffins referred to these rockets as 'jets'. JPL, the Jet Propulsion Laboratory, got its name in the same way.

The Arizona Highway Patrol were mystified when they came upon a pile of smoldering wreckage embedded in the side of a cliff rising above the road at the apex of a curve. The metal debris resembled the site of an airplane crash, but it turned out to be the vaporized remains of an automobile. The make of the vehicle was unidentifiable at the scene.

The folks in the lab finally figured out what it was, and pieced together the events that led up to its demise. It seems that a former Air Force sergeant had somehow got hold of a JATO (Jet Assisted Take-Off) unit. JATO units are solid fuel rockets used to give heavy military transport airplanes an extra push for take-off from short airfields.

Dried desert lakebeds are the location of choice for breaking the world ground vehicle speed record. The sergeant took the JATO unit into the Arizona desert and found a long, straight stretch of road. He attached the JATO unit to his car, jumped in, accelerated to a high speed, and fired off the rocket. The facts, as best as could be determined, are as follows:

The operator was driving a 1967 Chevy Impala. He ignited the JATO unit approximately 3.9 miles from the crash site. This was established by the location of a prominently scorched and melted strip of asphalt. The vehicle quickly reached a speed of between 250 and 300 mph and continued at that speed, under full power, for an additional 20–25 seconds. The soon-to-be pilot experienced G-forces usually reserved for dog-fighting F-14 jocks under full afterburners.

The Chevy remained on the straight highway for approximately 2.6 miles (15–20 seconds) before the driver applied the brakes, completely melting them, blowing the tyres, and leaving thick rubber marks on the road surface. The vehicle then became airborne for an additional 1.3 miles, impacted the cliff face at a height of 125 feet, and left a blackened crater 3 feet deep in the rock.

Most of the driver's remains were not recovered; however, small fragments of bone, teeth, and hair were extracted from the crater, and fingernail and bone shards were removed from a piece of debris believed to be a portion of the steering wheel.

Variants of this yarn seem to have been around since the 1960s, with other models of vehicle, but since about 1994, it has always been a 1967 Chevrolet Impala. The driver, incidentally, is often said to have been a Darwin award entrant, an honour granted to people who improve the gene pool by exiting it.

stopping distances

The distance you travel after you hit the brakes on a speeding vehicle will depend on many things. The most important is the vehicle's speed, followed by the effect of friction on the road. In the bigger picture, you also need to factor in reaction time. From the moment you see something until you react and hit the brakes will be around 0.7 seconds.

Imperial measure turns out to be more convenient for a quick calculation of the thinking-time distance: if you are going at 30 mph (48.2 km/h), you will travel 30 ft (9.1 m) before the brakes have an effect, at 60 mph (96.6 km/h), it will be 60 ft (18.3 m).

Figure 7. Stopping times and distance

Poor road, worn tyres				
km/h	mph	m	ft	seconds
30.0	18.6	15	48	4.2
60.0	37.3	47	154	6.3
90.0	55.9	97	317	8.4
120.0	74.6	164	540	10.5
150.0	93.2	250	820	12.7
180.0	111.9	353	1158	14.8
Fair road, reasonable tyres				
km/h	mph	m	ft	seconds
30.0	18.6	12	38	3.5
60.0	37.3	35	115	4.9
90.0	55.9	70	230	6.3
120.0	74.6	117	385	7.7
150.0	93.2	176	578	9.1
180.0	111.9	247	809	10.5
Good road, excellent tyres				
km/h	mph	m	ft	seconds
30.0	18.6	10	33	3.1
60.0	37.3	29	95	4.2
90.0	55.9	57	187	5.2
120.0	74.6	94	307	6.3
150.0	93.2	139	456	7.4
180.0	111.9	194	635	8.4

Figure 7 shows the stopping distances and times taken before stopping for poor, medium and good road and tyre conditions which influence the coefficient of friction (taken here as 0.4, 0.6 and 0.8 respectively).

Without going into the details of how it is arrived at, the distance travelled by an optimally braked vehicle is given by this formula, where the Greek letter µ (mu) is the coefficient of friction and g is the acceleration due to gravity: $s = v^2/2\mu g$

What happens when there are no brakes, when the engine is just turned off and the vehicle coasts to a stop? Recent Australian research on heavy vehicles over flat ground reveals that the *distance* travelled during a drop of 15 km/h (9.3 mph) does not depend on the initial speed, and that the distance travelled while shedding 30 km/h (18.6 mph) is twice the distance taken to lose 15 km/h (9.3 mph). The distance travelled does, however, vary with different sorts of heavy vehicle.

So going on that, the brakes are probably the way to go if you really want to stop!

the red flag act

Around 1769, the world's first self-propelled vehicle drove down a road in Paris. Designed by Nicolas Cugnot, it was a steam-powered military tractor which was going to haul artillery from place to place. It had a few problems, like the need to stop every 10 to 15 minutes to build up steam pressure again, and it only travelled at around 4 km/h (2.5 mph), a slow walking pace for a horse.

Slow though it may have been, another version of this slowly lumbering cart ran into (and demolished) a stone wall in 1771, and France went sour on such monstrous frivolities for a period. But by 1859,

Figure 8. Slowing distances

Type of vehicle	Distance to drop from 115 to 100 km/h	Distance to drop from 71 to 62 mph
Semi-trailer	705 m	771 yd
B-double	830 m	908 yd
Road train	1030 m	1126 yd
Coach	580 m	634 yd
Tipper and dog-trailer	650 m	711 yd

steam vehicles were being seriously considered, and some of them were designed to run on roads without the need for rails. In Britain, new laws were brought in to control these puffing leviathans. It is usual to blame the proprietors of stage coaches for the new laws, but by this point, railways had largely done away with those, and the laws were more to keep the new vehicles under control.

The Locomotive Act of 1861 set out the rules for tolls on turnpike roads: as rates usually related to the number of horses, locomotives were to pay on the basis that each 2 ton weight or part thereof was equal to one horse (for toll purposes only). This did not matter much, as most were just self-propelled agricultural steam engines, which never went far on the roads.

An 1865 amendment set a town speed limit of 2 mph (3.2 km/h) and a rural speed limit of 4 mph (6.4 km/h), and required a driver and a stoker on board, with a third person walking 60 yd (55 m) ahead with a red flag (a lantern at night), to warn oncoming traffic. The red flag was later dropped, but the person in front remained a requirement until the passing of the Locomotives on the Highway Act of 1896. In its final years, the Act restricted the operation of motor vehicles on English roads.

slow boats and fast boats on the mekong

Along Asia's Mekong River, where 'Red Flag' usually has a different connotation, boat travel is one of the more popular modes of transport. Canoes were originally used, when the time was right, but today canoes are only used for local travel, while longer-distance travel is by powered boats.

The Mekong is fed by meltwaters from as far away as the Tibetan Himalayas, almost 5000 km (3000 miles) from its mouth, and the annual flow is 475 km³ (114 cubic miles). There is a flood season roughly from May to November, and a dry season from December to April.

At places, the river is made impassable by falls, while rapids at other points put a limit on the boats that pass by. At high water, the currents may be too great, and at low water, there may be too many rocks exposed; for a short period early in the year, long sections of the river are navigable.

The first known long-distance expedition was the French Mekong Expedition taking two years to travel from the mouth of the river to Yunnan in China, with a few side excursions. Today's tourist adventurers travel between the Thai town of Chiang Khong, boarding their boats at Ban Houei Xai on the Lao side of the river and travelling down to Luang Prabang in Laos. This is about 250 km (155 miles) along the river, with 'slow boats' taking two days for the journey – and apparently they take the same time on return trips, against the current. Then there are the 'fast boats', sharp-looking canoes where the tourists wear crash helmets as they hurtle the distance in a few hours at about 40 knots, 75 km/h (46.6 mph).

A paddled canoe is quieter, but to get through rapids, you need engine power, so give me a 'slow boat', any day!

Figure 9. Fastest Atlantic crossings

Steamer	Year	City to destination	Nautical miles	Voyage-time (d-h-m)	Knots
Sirius	1838	New York to Falmouth	3159	18-00-00	7.31
Great Western	1839	Avonmouth to New York	3140	14-16-00	8.92
Great Western	1838	New York to Avonmouth	3218	14-15-59	9.14
Great Western	1843	Liverpool to New York	3068	12-18-00	10.03
Asia	1850	Liverpool to Halifax	2534	8-14-50	12.25
Pacific	1851	New York to Liverpool	3078	9-20-14	13.03
Persia	1856	Sandy Hook to Liverpool	3046	8-23-19	14.15
Adriatic	1872	Queenstown to Sandy Hook	2778	7-23-17	14.53
Baltic	1873	Sandy Hook to Queenstown	2840	7-20-09	15.09
Alaska	1883	Queenstown to Sandy Hook	2844	6-23-48	17.05
Oregon	1884	Sandy Hook to Queenstown	2861	6-16-57	18.09
City of Paris	1889	Queenstown to Sandy Hook	2788	5-19-18	20.01
Campania	1893	Queenstown to Sandy Hook	2864	5-15-37	21.12
Mauretania	1908	Sandy Hook to Queenstown	2932	5-00-05	24.42
Lusitania	1908	Queenstown to Sandy Hook	2891	4-19-36	25.01
Bremen	1929	Cherbourg to Ambrose	3164	4-17-42	27.83
Bremen	1933	Ambrose to Cherbourg	3199	4-16-15	28.51
Queen Mary	1936	Bishop Rock to Ambrose	2907	4-0-27	30.14
Normandie	1937	Ambrose to Bishop Rock	2936	3-22-07	31.2

the pursuit of the blue riband

In 1819, the *Savannah* took 29 days (some accounts say 21 days, but 29 is more likely) to cross the Atlantic from Savannah, Georgia to Liverpool in the UK. She only used steam for a few hours of that time. The *Royal William* was the next record holder, steaming over the Atlantic from Pictou to the Isle of Wight in 22 days in 1833. From 1838 onwards, the ship that made the fastest crossing was awarded the 'Blue Riband', a pennant to be flown from the topmast. Figure 9 on page 97 has a few of the highlights of the competition.

powerboat speed records

For the first century after Robert Fulton's pioneering 1802 steam boat, nobody really cared about the speed of powerboats. They were workhorses, used as tugs to pull ships away from their berths and to haul barges and canal boats at speeds faster than a horse, but the idea of a speedboat, a boat with an engine that could be raced, seems never to have emerged.

No doubt a few tugs raced each other, just as paddle-steamers raced on the Murray and the Mississippi rivers, but people did not head to the workshop or sit down to plan building a faster boat. That changed in 1908 when telephone inventor and engineer, Alexander Graham Bell, began experimenting with Casey Baldwin. In 1919, their hydrofoil set a water speed record of 70.86 mph (114.01 km/h).

During the 1920s, the water record stayed in the USA as speeds rose towards 100 mph (160 km/h). In 1930, the rules were changed, as they had been for land speed records. Now there had to be two runs in opposite directions, with the average of the two being taken. The same year saw a record of 98.76 mph (158.9 km/h) set on Britain's Lake

Windermere, but Sir Henry Segrave went back for a third run, in the hope of breaking 100 mph (160.9 km/h), and crashed, killing himself and his co-pilot in the process.

There is probably no more dangerous pursuit than seeking the water speed record, and it is estimated that the chances of death are somewhere above 50%. While the various attempts have no doubt helped develop the science of very fast watercraft, little benefit to humanity has come from them.

The present record was set by Ken Warby on Blowering Dam, Australia, in October 1978, with a speed of 317.596 mph (511.13 km/h). Since that time, there have been two official attempts on the record, and two fatalities.

sailing records

In a howling gale, you might think that a sailing ship would be pushed along as fast as the gale, but there are a few other factors to take into account. In the first place, big winds usually mean big seas, and as any small-boat sailor who has sailed in a swell can tell you, there is little wind when a vessel is in the trough of a wave.

A US Navy oiler, the USS *Ramapo*, encountered truly monumental waves in February 1933 in the northern Pacific. The ship survived, and brought back reports of waves measuring 34 m (112 ft) from crest to trough, enough depth to spill the wind from the sails of most vessels.

The oiler was small, only 146 m (479 ft) long, compared with the wavelength, the distance between crests, of 342 m (1122 ft), so it survived. Many modern tankers would span two crests, and with no support amidships, would be at risk of breaking. Ships simply do not have the rigidity of thousand-foot bridges.

As a rough approximation, waves in open water travel at a figure found by multiplying the square root of the wavelength by 1.25. With that equation, the waves the *Ramapo* encountered were travelling at about 23 m/s (75.5 ft/s), more than 80 km/h (50 mph). Under ideal conditions, that would probably be the limit to the speed of a ship, but any vessel travelling like that, at the whim of wind and waves, is unlikely to last long.

The ideal conditions for setting sailing speed records involve smooth water. In the case of powerboats, mirror-smooth water is needed, but vessels under sail must compromise, seeking high winds and water as smooth as they can get.

There are other limits as well: a sailing vessel only sails well because of the way the forces on the hull, keel and rudder under the water interact with forces on the hull, rigging and sails above the water. Conventional vessels need a deep keel for stability and to take a grip on the ocean, but vessels with multiple hulls can do away with this and skitter over the surface, which explains why catamarans and trimarans featured so prominently in sailing records in the twentieth century.

Today, records are kept by the World Sailing Speed Record Council, who report that a trimaran named *Groupama 3* crossed the Atlantic, west to east in July 2007 at an average speed of 29.26 knots, or 53.5 km/h (33 mph), and the same vessel holds several other records as well. The fastest monohull crossing at that time was *MariCha IV*, with 18.05 knots, or 33 km/h (20.5 mph). The monohull speed would have been good enough to take the Blue Riband for much of the nineteenth century, but *Groupama 3* would have held it at any time before 1936.

While they are not suited for ocean crossings, windsurfers are also incredibly fast. The record for a man over a nautical mile is held by Bjorn Dunkerbeck, who achieved 41.14 knots in 2006: he first held the record in 2003 with 33.96 knots, when Britt Dunkerbeck (his sister) achieved a women's record of 23.84 knots. That record now stands at 34.74 knots, sailed by Zara Davis in 2006.

The nautical-mile speed record depends on luck, to a certain extent, but the distance sailed in a day is a better measure of sustained effort. In 1854, a clipper named *Champion of the Sea* completed 467 nmi in a day at 19.46 knots. In 1984, a trimaran named *Formule Tag* sailed 512.5 nmi at 21.35 knots. *Groupama 3* stars in this category as well, with 794 nmi in a day, at an average of 33.08 knots or 60.5 km/h (37.5 mph). Among single-handed craft, *Brossard* sailed 610.45 nmi in 2006 at an average of 25.76 knots or 47.1 km/h (29.3 mph).

how speed shaped our society

Our modern world owes its existence to speed. The speed that carried copper ore from Cuba to Wales, the speed that carried gutta percha, a rubbery substance used as insulation, from Malaya to Europe, the speed that carried the finished copper cable to Sardinia, where a cable under the sea could improve communications; all of these changed the way the world lived and thought.

India rubber from the Amazon might be combined with cotton from Louisiana to make a belt that would be rushed around the world to transfer steam power to machinery in Australia or to make a fire hose that would be used in Moscow. Peace on the high seas and speed were the keys to these innovations.

The big change with speed and our society was with the cities and their swelling suburbs. Horse-drawn trams and omnibuses carried people shorter distances, but by the 1840s, cities were approaching gridlock. By the 1860s, serious attempts were being made to move railways underground, to hurry people across town or out of town.

With fast transport in place, the city could remain the centre of activity and commerce (and for a while, the location of the belching mills, factories and forges). Some people, the better-off, no longer had a need to live in the city, amidst the foul air that was popularly believed to cause illness, and the noise and bustle. Instead, people could live in suburbs with lawns and gardens, and still get to their place of work and back each day.

They were also divorced from the smell and muck that is associated with growing and making food. Fast trains could carry food to the city and into the suburbs each night, ready to be sold the next morning. Speed made the cities, and enslaved the male to his lawnmower.

tea clippers and fast passages

In England, at least, if speed and steam trains allowed the cities to exist, it was tea that kept them alive. The big killer of the crowded nineteenth-century cities was cholera, and for most of the century, tea drinking kept the British public alive. They survived in blissful ignorance of tea's usefulness, because it was not so much the act of drinking tea, as the boiling of the water that did the good.

Tea in 1850 was produced mainly in China, and there was a firm belief that the new season's tea always tasted better. This is a reasonable view to take, because at the end of a season, there would be more adulterated tea – iron sulfate, calcium sulfate, turmeric, soapstone and Prussian blue were all used to make the tea seem fresher.

Tea merchants saw that the first of a new season's tea could be sold at a premium, but only those who got their hands on the crop first would reap the benefits. Fast ships were needed, and the fastest were sleek 'Baltimore clippers', topsail schooners which sailed close to the wind and reached port sooner. Soon after 1850, the yards started turning out tea clippers which could rush their precious cargoes to market. The era of the tea clipper lasted barely two decades before sturdy steamers, able to push straight into the wind, took over.

Clippers sailed at 9 knots, or 17 km/h (10.5 mph), many achieving 20 knots, or 37 km/h (23 mph) at sea. One ship, *Sovereign of the Seas*, logged 22 knots, or 41 km/h (25 mph) down the Australian coast in 1854, and it was not unheard-of for ships to exceed 400 nmi, or 750 km (466 miles) in a day. Only one of those ships survives to this day, *Cutty Sark*, land-locked at Greenwich near London, and unfortunately badly damaged in a fire in 2007.

fast transport and slow deaths

The delivery of tea to England by fast ship might have kept the English boiling water to make tea, killing the bacteria that caused cholera and other diseases. On the other hand, fast transport also caused some curious outbreaks of other diseases.

Airport malaria is a known phenomenon today, where people close to airports can catch malaria when an infected mosquito emerges from an aircraft and draws blood from somebody before it dies. That sort of thing was far less likely in the days of steamships, although not impossible, in the case of the barque *Hecla*, which once carried yellow fever to Wales – and she was a sailing ship!

Hecla reached Swansea with a cargo of copper ore from Cuba on 8 September 1865, under-crewed due to deaths that were put down to 'dropsy', but which were almost certainly yellow fever. James Saunders, another sailor, died just after landing, and judged to have been a victim of yellow fever, his body was immediately buried in a tar sheet, his house was cleared and disinfected with lime wash and chloride of lime, and his clothing and bedding were destroyed.

Nobody had any idea that the disease was spread by mosquito bites, so the ship's water supply, almost certainly complete with mosquitoes in all stages of life, was left unexamined. While the ship was disinfected and later moved when locals intimated that it might mysteriously catch fire, neither would have had an effect on stopping the mosquitoes.

Some 29 people fell ill with yellow fever and 17 died, with a few other 'possibles'. The ship had travelled in warm weather, allowing the mosquitoes to survive, and arrived in warm weather, which allowed them to spread, briefly, into parts of Swansea. Yellow fever and its mosquitoes had travelled to the Caribbean with African slave ships and been established there, but Wales was safe from any permanent threat from yellow fever, back then. In these days of global warming, who can say what the future might hold?

While political unrest was common in China in the late nineteenth century, new technology brought hope to some of China's urban poor. They could take steam trains into rural areas to shoot, kill and skin ground rodents, and then take the skins back into the city for sale. In an age before plastics, skins were always saleable, and while the local country people had silly traditions, like not shooting a sick-looking animal, the city slickers saw them all as fair and easy game. They were wrong.

Bubonic plague is a disease that harms rodents, fleas and humans. When a flea bites an infected mammal, it gets an infection that blocks its blood-sucking apparatus. The next time it tries to feed, some of the plague bacteria are 'blown back' into the new food source, and so the disease spreads. When a host dies, fleas move to any other warm body – the person skinning the old host, for example.

The hunters caught fast steam trains back to the city before they fell ill, and from there, bubonic plague infected rats in the city, either from the hunters, or from the fleas themselves that were still in the fur of the skins. Over time, some of the rats found their way onto fast steamships that left Chinese ports and sailed around the world.

In earlier times, the plague usually killed the rats before sailing ships reached port, but as steamships bustled from port to port, sooner or later some of the rats reached the other end, made it ashore, and shared their fleas and their ills. Indian ports were hit, along with Sydney, San Francisco, Madagascar, Paraguay, South Africa and more. In every port, people died because of fast ships.

records with the oar

When Vikings sailed the oceans, they could use sails, but for the most part, they relied on oars. Most naval engagements fought in the Mediterranean until 1500 involved boats that were driven by oars: triremes, quinqueremes, Moorish galleys and more. Sails and wind just weren't reliable enough, because if the wind dropped or came from the wrong direction, only its rowers could move the ship.

In the same way, those who put to sea to fish for the most part used oars; whales were even harpooned from boats that were rowed. Of course, if the harpoon went home, the boat crew might be treated to a 'Nantucket sleigh ride' as the whale plunged, dived and fought against the pain of its wound. Then it was back to the oars again, to haul the carcass back to the ship.

In 1896, two Norwegians left Battery Park at the foot of Manhattan, and set out for Europe in an open boat. George Harbo and Frank Samuelson reached the Isles of Scilly, 55 days later. Since that time, some 177 other rowboats have successfully crossed the Atlantic, 68 of them with just a single rower, taking between 2 and 6 months. The fastest solo row so far is 42 days, and a team of four once made it in 36 days.

A French team of eleven crossed from the Canary Islands to Martinique in 1992, a distance of 2565 nmi, or 4750 km (2956 miles) in 35 days, 8 hours and 30 minutes, an average of 3.02 knots, or 5.6 km/h (3.5 mph). By comparison, the Oxford-Cambridge boat race is 4 miles, 374 yards (6.779 km), for which the record, set by Cambridge in 1998 is 16 minutes, 19 seconds. This is an average of 13.6 knots, or 24.92 km/h, (15.49 mph). Mind you, the French team's *Mondiale* had eight rowing and six sleeping positions, and they started out with· 3¼ tons of supplies.

around the world in 80 ways

When Ferdinand Magellan's expedition crawled back home, the world's first recorded circumnavigation had lasted 1083 days. In the years that followed, a number of trips would exceed that time – the voyage of the *Beagle* for example took almost 5 years, but that was because Mr Darwin kept stopping off to poke at fossils and take specimens, and Captain FitzRoy stopped to make maps. For the most part, as time has gone by, travelling around the globe has always become faster and faster.

In 1860, a New Zealander could travel by steamer to Sydney, then by steamer to Suez (where canal-digging had just started), train to Alexandria, steamer to Marseilles, train to Calais, steamer to Dover, and train to London in 80 days. If connections were available, the return journey by steamer to Panama, train to the Pacific and steamer to New Zealand would take even less time.

Over a period of 5 years in the 1930s, the widely admired *Graf Zeppelin* travelled 743,248 km (461,833 miles) in 7342 hours, 51 minutes, an average of 101.2 km/h (62.9 mph). At that speed, nonstop, it would have taken the airship 16 days and 13 hours to go around the world.

Even in 1924, a 'flying boat' could carry six passengers at 136.8 km/h (85 mph), but many stops were required, so progress was slow. In 1931, Wiley Post and Harold Gatty took a short-course round-the-world trip of some 24,000 km (15,000 miles) in 8 days and 15 hours, then Post flew the same route solo in 1934 in 7 days, 19 hours, logging averages of 120.2 km/h (74.7 mph) the first time and 133.1 km/h (82.7 mph) the second.

By 1932, Imperial Airways was carrying passengers at 167 km/h (105 mph), and aircraft were reaching further. In 1939, the Dutch airline KLM assured New Guinea planters that after a steamer to Sydney, they could reach Amsterdam by modern aircraft in just 7 days, a time which included regular stops.

the fossett phenomenon

On 12 July 2007, Steve Fossett, from the United States and Terry Delore from New Zealand flew a glider around a 1250 km (777 mile) triangular course in Nevada in 8 hours and 23 minutes, at a new world record average speed of 149.23 km/h (92.73 mph).

In doing so, they eclipsed the last of the many records set by German Hans Werner Grosse, who set his record in the Alice Springs area of Australia in January 1987. Grosse had achieved a speed for the triangular course of 143.46 km/h (89.14 mph). In their pursuit of the elusive record, the pilots had made attempts over 4 years in Argentina, Australia, South Africa and the United States.

In 2006, Fossett flew solo in the Virgin Atlantic GlobalFlyer to set both the Absolute Non-Stop Distance Record for any aircraft (February 2006) and the Absolute Closed Circuit Distance Record (March 2006) as well as establishing a new glider world altitude record (August 2006) with co-pilot Einar Enevoldson. Not surprisingly, he was added to the USA's National Aviation Hall of Fame on 21 July 2007, in recognition of his world record achievements in four categories of aircraft: gliders, balloons, aeroplanes and airships. After their triangular course record, in July, Delore and Fossett said they planned to set more glider records together in November 2007 in Argentina.

Fossett also hoped to break 800 mph (1287 km/h) on land in the S&S LM-1500 turbo-jet-powered racer that began as the third of Craig Breedlove's 'Spirit of America' series challengers. It was originally designed to break Richard Noble's 1983 record of 633 mph (1,019 km/h), but Fossett took it on, buying the project from Breedlove in mid-2006.

In September, while seeking a better lake-bed for the land speed attempt, Fossett's blue and white plane disappeared, somewhere in Nevada. Searchers found half a dozen other crash sites, but not Fossett's. Although his fate remains a mystery, he was declared legally dead in early 2008.

the world solar challenge

The 'Challenge' is a race in which solar-powered cars race 3000 km (about 2000 miles) from Darwin, in the north of Australia, to Adelaide, in the south, with some very strict rules to abide by. It used to be held every three years, but since 1999, has been held every two years.

The drivers must be at least 80 kg (176 lb), or else compensating ballast must be added, and cars may only race during specified hours: they can roll on for a little past the finishing time to a convenient stopping-place, but they must start again the next day from where they were at 5 pm the previous day.

The first race, in 1987, was won in 44 hours and 54 minutes at an average speed of 66.9 km/h (41.6 mph). In 1990, the winner took 46 hours and 8 minutes, an average of 65.18 km/h (40.5 mph). In 1993, Team Honda took 35 hours and 28 minutes (84.95 km/h, or 52.8 mph), while the 1996 race was won by the same team in 33 hours and 32 minutes, an average of 89.76 km/h (55.8 mph).

In 1999, the winning time was set by an Australian team in 41 hours and 6 minutes (72.96 km/h, or 45.3 mph), while in 2001 and 2003 the Dutch team Nuna won, in 32 hours, 39 minutes (91.81 km/h, or 57 mph), and 30 hours and 54 minutes (97.02 km/h, or 60.3 mph) respectively. In 2005, they won again, in 29 hours and 11 minutes, an average of 103.5 km/h (64.3 mph).

Thinking had to be changed for the 2007 race, because vehicles were now travelling so fast that they were affected by the speed limit that applied to the roads once they crossed from the Northern Territory (where there is no limit) into South Australia – now even the northern section has a 130 km/h (81 mph) limit. Support vehicles also had trouble keeping up, and so the rules were changed for the new Challenge class. The old competitors looked like a cross between an ironing board and a cockroach, but future vehicles will look more normal, and will carry fewer solar cells.

a fast train

On the banks of the Thames near Chelsea, there is a statue of a British member of Parliament, William Huskisson, shown clad in a toga. There is no reason given for the statue being there, no plaque to indicate why he stands beside the Thames, but I think I at least know why the statue was created. In 1830, railways claimed their first pedestrian victim when

Mr Huskisson wandered onto a track, and failed to get out of the way as a famous locomotive, George Stephenson's *Rocket*, came rushing through at 50 km/h (30 mph).

Mr Huskisson died because he had no sense of speed, no way of realising how quickly the locomotive would arrive. Coaches travelled at perhaps 16 km/h (10 mph), and nothing on wheels went any faster. Lady Wilton, an eyewitness, described how the MP lost one part, figuratively, and another part, quite literally:

> ... poor Mr. Huskisson, less active from the effects of age and ill-health, bewildered, too, by the frantic cries of 'Stop the engine! Clear the track!' that resounded on all sides, completely lost his head, looked helplessly to the right and left, and was instantaneously prostrated by the fatal machine, which dashed down like a thunderbolt upon him, and passed over his leg, smashing and mangling it in the most horrible way.

Stephenson's *Rocket* was one of several locomotives on the world's first commercial steam railway, but it would be a mistake to think that this was the first-ever railway: rails had been used for some time for horse-drawn loads. A horse could carry 135 kg (300 lb), haul 270 kg (600 lb) in a cart, 2 tons in a cart on rails, or 20 tons in a canal barge.

Canals were used for long distances, but for shorter distances, horse railways were a 'big thing' in the early nineteenth century, both for passengers in cities and transferring heavy goods to other transport.

Up until the late 1850s, as trains spread across the world, a speed of 50 km/h (30 mph) seemed good. After all, trains could go most of the day and most of the night, with just a few stops to set down or pick up, or to take on fuel and water. There were also occasional comfort breaks, because for many years, there were no toilets on trains, although one could always use a chamber pot and empty it sneakily when the train slowed. After 1860, sleeping cars were introduced. At 50 km/h (30 mph), a train could cover 1000 km (600 miles) or more in a day and a night.

By this point, a few trains were going faster than 50 km/h (30 mph): by 1850, *Great Britain* was claimed to have reached 130 km/h (80 mph) and another steam locomotive on the Bristol and Exeter Railway claimed slightly more. Up until 1860 though, most were limited to 50 km/h (30 mph). Engineers debated whether it was the track or the locomotives that kept trains slow, but it was more to do with lubrication: fast engines needed oil, not grease, and it was hard to find the right oils until people started sinking oil wells.

By the early 1900s, engineers started to understand what they were doing, and a first, a train travelling at 160.9 km/h (100 mph), was claimed for a GWR locomotive, *City of Truro* in 1904. The first authenticated railway 'ton' wasn't until the *Flying Scotsman* in 1934. Speeds of 200 km/h (125 mph) were exceeded in both Britain and Germany in 1936 and 1938, but the age of steam was passing.

The top speed of the French Train Grand Vitesse (TGV – or very fast train) in 1990 was 507.6 km/h (321 mph), but by 2007, it eclipsed its previous record with 574.8 km/h (357.3 mph). The Japanese Shinkansen, or Bullet Train, started out in 1964 at 210 km/h (130 mph), but commonly achieves 300 km/h (188 mph), while conventional rail sets have clocked 443 km/h (275 mph) with a world record set by Maglev train sets in 2003 at 581 km/h (361 mph).

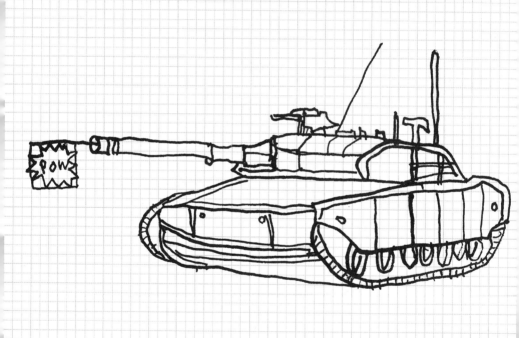

on the warpath

To win a war, you have to be fast – right? Well, maybe … but what about sneaking up on the enemy? Sometimes slow and sneaky works just as well.

In World War II, the word 'blitz' came into the English language. Nowadays, we more commonly hear the term used in relation to a backyard or a bedroom being 'blitzed', being turned over and renewed. Originally, it referred to the German *blitzkrieg*, or lightning war, where a massive force was brought to bear, using mechanised warfare to move troops from place to place fast. Any place hit by a *blitzkrieg* was certainly in need of renewal.

Blitzkrieg worked well at first, that is, until the German command turned away from the British troops who were trapped near Dunkirk. Paris seemed too attractive not to take it first, and so what if a few Allied troops escaped?

It turned out that more than just a few escaped, proving costly later. Reliance on *blitzkrieg* ideas also proved unwise when Germany attacked Russia (the USSR), and did not manage to finish the strike before winter set in.

The tank wars of northern Africa and Europe, however, showed how important speed could be, as did the Battle of Britain. It is hard to predict how important speed will be in any future war, but we can certainly look back and see how important it has been in the past.

charge!

According to Lord Cardigan, the Light Brigade charged the Russian guns in the Crimean War at 17 mph (27.4 km/h), a speed that was quite possible, though we have no idea how exactly he arrived at that particular figure.

Horses are herd animals, and so they are comfortable galloping together at close quarters. The idea of the cavalry charge was to form a line, with the horses close together, and then move forward, first at a trot, and later, over the last hundred yards (90 m), at full gallop, riding down a wretched infantry, terrifying them and causing them to flee.

The theory of the cavalry charge started to come apart when the longbow came into use. Long-bowmen could shower the approaching line with arrows, injuring the horses and possibly the riders, long before they could harm the infantry.

Then there were the pikemen, who used long pointy sticks to impale the horses as the charge reached them, the palisades and other defences. The cavalry could often manoeuvre around the defences, which were hard to shift around on the battlefield. Muskets on the other hand had a practical range of perhaps 90 m (100 yd), but even the best infantryman was unable to fire more than four shots a minute.

Allowing a reload time of 15 seconds, Cardigan's men would have covered 115 m (125 yd) between shots, so the Russian infantry had just one chance to kill, wound or stop the cavalry who were hurtling at them, half a ton of thundering horse with a razor-sharp sabre swinging down at their heads and shoulders.

Instead of attacking infantry, Cardigan found himself charging 50 guns that fired grape and round shot over far greater distances, with Russian riflemen on his flanks picking off more riders. The wonder is not that so many were killed (118) or wounded (127) from the 670 men who started out, but that so few were hit.

The tradition of the charge was maintained into World War I, when infantry were repeatedly sent 'over the top' to charge an enemy equipped with machine guns and repeating rifles with a far greater range. The Australian Light Horse undertook a successful cavalry charge, one of the very last, at Beersheba in 1917.

The Australian Light Horse were really mounted infantry who rode to the battle and dismounted to fight. Only Poland had cavalry in World War II, but they were mounted infantry like the Light Horse. The role of the cavalry passed to the much faster tank regiments, though cavalry ranks like squadron leader (major) and wing commander (lieutenant-colonel) in some air forces reflect their cavalry origins. Tank regiment officers often wear silver insignia, another cavalry tradition.

Aircraft were obviously never used to charge infantry, although they did provide air cover at times to the tanks that were often used, like cavalry, for infantry support in World War II. In this day and age, the charge has largely ceased to be an effective military tactic.

Animals still charge, though. An elephant at 50 km/h (31 mph) weighs as much as 12 horses and travels roughly twice as fast. American bison weigh about a ton, but have the same top speed. The bulls of Pamplona, the ones that people run through the streets with, have been clocked at 55 km/h (34 mph), the same speed reported for giraffes and African buffalo. More on these later.

the moose cavalry threat

Because horses are herd animals, seen as food by larger predators, they get nervous easily. A wild stallion may move forward to threaten a human, but picking up a stick is enough to make the horse think again, and retreat.

Clearly, horses are fairly easily 'spooked', and people who have spent some time around horses are well aware of this. They will also assure anybody who will listen that the smell of a strange animal will always upset a horse. Camels and Hannibal's elephants have both been cited in the past as causing stampedes of cavalry horses.

The moose is another animal that has been used for 'cavalry', if the tales are to be believed. They were used in a number of northern European cultures as a draught animal, pulling sleighs, and even carrying riders on occasion.

Blame Monty Python, who made the moose a running joke, for some of the confusion, but there were others as well. One hoaxer has produced an excellently manipulated digital image of a harnessed moose, supposedly used to haul logs out of forests. The giveaway is that an identical pile of logs appears on each side of the picture, one of the elements reversed – if there has been one legpull, there may have been other hoaxes.

You can read that Ivan the Terrible banned moose husbandry in Siberia in the 1500s to stop the locals using moose cavalry against him. Or that in the 1700s, Sweden tried moose cavalry, because they believed the moose smell would terrify enemy horses – the idea supposedly failed when it proved too hard to gather food for the animals, which seems improbable, given that they eat twigs. Forces of moose cavalry may or may not have existed, but if they ever did, a moose cavalry would have charged at 55 km/h (34 mph) – and provided the Python team with some excellent material.

slow wars

Barring a nuclear war, history is unlikely to count the 6-day war between Egypt and Israel as a slow war, but compared with some of the other wars of history, it was pretty fast.

The champion of slow wars was probably the Hundred Years War, which really lasted 116 years, from 1337 to 1453. Between England and France, it was fought over who was to rule France. In the end, the British were forced out of Calais, and that was that. Between 1689 and 1815 though, France and Britain were nearly always at war, and the 126-year period is sometimes called the 'second hundred years war'.

The Thirty Years War began in 1618 and ended in 1648. It was a religious war between Catholic nations and Protestant ones — though Catholic France saw fit at one point to side with the Protestants, in order to limit the power of the Habsburgs. At least this one had the grace to be fought mainly in Europe.

The Seven Years War took place between 1756 and 1763. It was a world war, with fighting in North America, Europe and Asia. There were other wars of the same name: in 1563–70 a number of nations were involved in fighting Sweden; and in Korea, a conflict between Korea and Japan took place from 1592–98.

World War II, at 6 years (1939–45), was one of the longer wars, while World War I, which seemed to go on forever, lasted just over 4 years. By way of comparison, the first Vietnam or Indochina War went from about 1945 to 1954, while what Westerners usually call the Vietnam War is generally reckoned to have started in 1959 and ended in 1975. Still, each of those is a pup compared with the Korean War, which saw hostilities from 1950 to 1953, but which, at the time of writing, had still not officially ended.

forced marches in war

In military history, there have only been a few cases where large forces were moved quickly over large distances, but these were not often successful. One of the last, Napoleon's 1812 advance on Moscow, was swift and effective but led to a later disastrous retreat.

Looking back, Hannibal was one of the success stories, with a most effective attack on Rome, based on a 15-day forced march over the Alps into the north of Italy in 218 BC. This was the famous crossing-with-elephants, and Hannibal's forces moved so fast that they almost caught the Romans by surprise. Hannibal won repeated victories after that, and it took Rome another 16 years to finally beat him after having lost the initiative.

Julius Caesar used forced marches very effectively against the enemy. On the other hand, his rival Marcus Licinius Crassus did not. He invaded Parthia, more or less where Turkey is today, in 53 BC. He was given false advice and, tricked into a forced march, chased a Parthian army near Carrhae until his army was strung out into a long line of marching infantry in a hot desert. Then the Parthian cavalry and horse-archers swept in, and defeated the much larger Roman army.

A later Roman leader, the emperor Heraclius, was far more successful, for a while, with forced marches. In 615 AD, Chosroes II of Persia held most of the Middle East. When an unknown citizen of Mecca, Muhammad, tried to form an alliance with Chosroes, the Persian emperor rejected him. The Persians were winning a long war against the forces of Byzantium (Constantinople), ruled by Heraclius at the time.

In 622, Muhammad moved to Medina just as Heraclius led his troops on a campaign that included 77 km (48 mile) marches in 24 hours. In a short time, Heraclius out-fought and out-thought the Persians and tore apart their empire. He forced Chosroes into defeat after defeat and retreat after retreat, until in 627, Chosroes was deposed. He died in a dungeon, five days later.

Rome regained all of its lost territory, but within a few years, the weakened Persian and Roman empires fell to the forces that had flocked to the once-obscure upstart Muhammad. Byzantium and Persia had worn each other out and, in the last 8 years of his rule, Heraclius saw all of his regained provinces fall to the forces of Islam.

Further north, in September of 1066, the Saxons were bracing for an invasion by the Normans when Harold Hardrada of Norway landed in the Humber estuary and moved on York in the north of England. The English king, Harold Godwinson, took his troops on a forced march and defeated the Viking forces at Stamford Bridge on 25 September. Three days later, the Normans came ashore in the south of England.

Hearing of this, Harold marched his troops the 320 km (200 miles) back to London between 3 October and 6 October. They rested briefly in London before pushing on another 80 km (50 miles) to Hastings, arriving on 13 October. They had covered 400 km (250 miles), in 10 days.

The two armies each had around 7000 men, but the Saxons were tired, and the Norman cavalry were superior. While it could have gone either way, in the end, the tiredness from the long march played a role and the Normans came out on top.

In 1935, in what seems to have been the last successful forced march, a Red Chinese battalion completed a night march of 135 km (85 miles) to descend on and capture Chou P'ing fort. They were wearing captured Nanking uniforms and they walked right in to the fort. The rest of the Red Army followed at a steadier pace.

Forced marches may still have a place in guerrilla warfare, but as transport methods get better and as aerial and satellite surveillance get better, it becomes harder for guerrillas to succeed against a highly organised and mechanised army. The art of moving an army fast has changed.

fetchez la vache

Before the invention of the cannon, you could defend yourselves from enemy attack by piling up earth, wood or rock, and sheltering behind it. The enemy would then do their best to knock your pile down, but charging in with a battering ram was a fairly unpleasant business. The defending side would drop rocks, the contents of the slops buckets, boiling oil (or worse) on you.

The walls of castles and cities were vertical to make them harder to climb. For the most part, the stones were not cemented, so if they were hit with a heavy enough weight, there was a chance they could fall down. Jumping ahead in time, once the cannon came into play, it was possible to bang away at the bottom of a wall, knock out a few stones, and bring down a whole section. People learned from this and sloped the walls, which deflected the cannon balls upwards.

In pre-cannon days, to get things over the walls of castles and cities, there was always the trebuchet, the mangonel or the onager, all styles of catapult that threw rocks from a safe distance. They could be used from far enough away that the operators were safe from boiling oil and slops buckets, and fairly safe from arrows – let's say a distance of about 150 m (164 yd).

The bare minimum speed to just reach the base of the wall from a distance of 150 m (164 yd) is 27 m/s (29.5 yd/s), or 97.6 km/h (60.5 mph). Realistically, the arm is going to have to sling those rocks at speeds closer to 150 km/h (93 mph). In its day, apart from arrows, there weren't many things that travelled as fast as a rock hurled by a siege engine.

In *Monty Python and the Holy Grail*, the French are shown firing a dead cow at King Arthur from a ballista, and there are legends of the corpses of soldiers who had died of the plague being fired into besieged medieval towns, but there is no hard evidence for this. There is just a nasty suspicion, but chances are if they could've, they most definitely would've!

speeding arrows

Until about 1600, most military firepower, aside from the odd cannon used to batter walls from a distance, came from bows and arrows. In reality, up until the mid-1800s, it would have made more sense to keep using archers, because a skilled bowman could fire more shots faster, and do greater harm at the end of their range than a soldier equipped with a rifle or a musket.

Firearms got the nod because an archer had to be skilled, and those who used longbows had to be strong. On the other hand, the skill and strength needed to fire a crossbow were low, like those needed to discharge a firearm, but guns won out.

To consider the longbow first: in October 1415, the small English army of Henry V, about 6000 men, was faced at Agincourt with an army of 50,000 French men. The difference was not as great as you might think, because some 5000 of the Englishmen were skilled archers. The French army was mainly composed of cavalry, but in the face of a rain of arrows, the French cavalry were turned back on the French infantry, causing confusion that is generally bad for winning a battle.

A good archer could send off 10 arrows a minute, each of them leaving the bow at 60 m/s (65.5 yd/s) (more than 200 km/h, or 124 mph), and arriving a few seconds later, still carrying three-quarters of that speed. All of these are estimates, of course, but we do know that in 1590, Sir Roger Williams complained that only 10% of archers could do harm '12 or 14 score off', which is at 220 to 260 m (240 to 280 yd). Even at Waterloo in 1815, muskets had a range of less than 90 m (100 yd).

Much of the armour used at Agincourt was thin metal, about 1 mm (0.04 inch) thick – tests have shown that arrows would go through that. Some armour was up to 4 mm (0.16 inch) thick, which would have withstood arrows, but not crossbow bolts.

The crossbow has the advantage that it can be loaded in advance, and used when necessary. More importantly, it fires a heavy bolt with real killing power, and no real training is needed to use one, because the operation is intuitive: point, steady the bow and shoot. After a ranging shot or two, most operators can be accurate enough to be dangerous.

Most comparisons of the longbow and the crossbow rely on modern crossbows, using modern materials and throwing bolts much faster. A longbow was 2 m (6.5 ft) long, but the lath (the springy part) on a crossbow was generally a great deal shorter. It is likely that the earlier crossbow bolts were slower, but later ones are credited with ranges of 400 m (437 yd) and more. Allowing for air resistance, the bolts must have reached at least 75 m/s (83 yd/s), close to 300 km/h (186.5 mph).

The rate of fire of the crossbow was comparable to that of a trained musket user, with less chance of a misfire, making the changeover to firearms a bit odd.

the big guns of world war one

As the world prepared for the war, known as the Great War, the War to End All Wars or World War I, armaments manufacturers everywhere grew rich as they raced to produce bigger and better weapons. In Germany, the greatest builder and profit-taker was Alfred Krupp.

Krupp had been making a howitzer (a type of gun) since 1900, an artillery weapon that could hurl a 350 mm (13.8 inch) 400 kg (882 lb) shell over 10 km (6 miles). By 1912, he had a gun that would fire a 420 mm (16.5 inch) 1 ton shell 16 km (10 miles). The big drawback of his newer weapon was the size of the gun, which weighed 175 tons. It had to be broken down into five parts, that could only then be hauled by railway to where the gun was needed.

By 1914, a smaller gun, dubbed 'Big Bertha', after Frau Krupp, Bertha Krupp von Bohlen, was available. It weighed a mere 43 tons, and threw a 420 mm (16.5 inch) 1 ton shell over the same distance. This gun was small enough to be hauled around by tractors, and did tremendous damage during the war.

There is a simple rule of thumb for assessing the velocity of a projectile, assuming that the gun is pointed at 45° and ignoring air resistance. (In other words, watch out for spherical cows!) This will give us an indication: range = v^2/g, where g is the acceleration due to gravity (9.8 ms^{-2}, or 32 ft/sec/sec), and v is in m/s, or ft/s. A note of caution here: the horizontal component and the vertical component are equal, and each will be ($v/\sqrt{2}$).

If we take the range as 16,000 m (10 miles), we need the square root of the product of the range and g or $\sqrt{(16,000 \times 9.8)}$, which is 396 m/s (1299 ft/s) (1425 km/h, or 885.5 mph). The commonly quoted figures of 400 or 425 m/s (1312 or 1394 ft/s) are probably reasonable.

We also have the recollections of Major Wesener to consider. He reported that in August 1914, the shell took 60 seconds to travel a 4000 m (2.5 mile) range on a high trajectory. This would have delivered a plunging shot in about that time if the gun were aimed at about 80° from the horizontal. The shell would drop in at supersonic speed, blasting the target area without warning.

Later in the war, the Germans used the Paris Gun (often wrongly referred to as 'Big Bertha') to drop shells on Paris from 120 km (75 miles). Known to the Germans as *Kaiser Wilhelm Geschütz*, the Emperor William Gun, this was destroyed towards the end of the war, so the accounts are often contradictory.

The gun fired 94 kg (207 lb) shells 120 km (74.5 miles) into the city of Paris, and the aim was one of terrorism more than anything else. Big Bertha's shells destroyed fortifications, while the Paris Gun was designed to drop shells without warning on a civilian population and intended to damage morale. No sound was heard from the far-off gun, and these shells also came in at supersonic speed, giving no warning before they landed. For some time, French authorities suspected that they were bombs from a high-flying Zeppelin.

The muzzle velocity of the gun was 1600 m/s (5249 ft/s), a mile a second, and the shells wore out the inside of the barrel, requiring each shell to be slightly larger than the one before. Around 350 shells were fired, each shell killing an average of one Parisian and wounding two more.

At the time, the gun fired shells 40 km (25 miles) up into the outer reaches of the atmosphere at almost Mach 5. Until the first flights of the V2, these shells were the highest and fastest objects made by human hands.

the V1 and the V2

In World War II, German forces could get as close as 40 km (25 miles) to parts of England, well within range of any new Paris Gun. The Germans were aware though that the gun had not been effective in World War I, that bombing had come a long way in the two decades, and that a large gun would be vulnerable to bombers.

With so few targets in reach, and a severe risk of the gun being destroyed, the Germans needed a weapon with a greater range and a larger payload, something which did not need a giant and very visible gun, and so the V1 and V2 were conceived.

These two weapons each carried about the same amount of destructive force as the Big Bertha shells, 1 ton of high explosive. The difference was in the cost of construction: you could build a hundred V1s for the cost of a V2, yet Hitler's Germany kept on pouring scarce resources into the V2.

Both the V1 and the V2 got their first letter from the German word *Vergeltungswaffen*, meaning 'revenge weapon', and while each carried the same amount of explosive, they had little in common. The V2 was far more effective as a terror weapon because it was undetectable and unstoppable. It was simply launched into space (literally) after which it fell back to its target at supersonic speed, so the first inkling it was coming was when it exploded.

The V1 on the other hand was a pilot-less pulse jet, flying at 2000 m (1.24 miles) and at a conventional speed. It could be chased by fighters and shot down, or nudged out of the way, and it could be brought down by anti-aircraft fire. In the last 4 weeks that it was used, 67% of the V1s were destroyed. The V1 had a primitive control system that sent it on a pre-set path until it ran out of fuel. When the characteristic sound of its 'buzz-bomb' engine stopped, people knew to take cover. Because the British controlled many German agents in Britain, they could provide misleading information about where the V1 bombs landed, and so cause more of the later ones to be sent in the wrong direction.

Like the Paris Gun, the V1 and V2 weapons could not be directed at a particular target, so they were mainly used to attack cities and civilians in an attempt to break morale. One thing you will rarely find in English-language accounts is that twice as many V2s were fired at Antwerp as were fired at London. This happened after D-Day, when Antwerp was being used as a major port for landing war materials. In one blanket attack, a large number of rockets were fired, causing a great deal of damage to the docks. One rocket hit an Antwerp cinema, killing 567 people.

The V2s killed approximately 7000 people when they landed, and 20,000 slave labourers died making them. The rocket motor carried it to 80 km (50 miles), after which it was a ballistic missile, operating without force or control. It came in at about Mach 4, with no sound before the explosion which was followed by an eerie whistling sound, after the bang.

The British first got wind of the V2 plans in 1939, 3 years before the rocket flew. For a long while, their best estimate of the warhead's mass was 10 tons, more than enough capacity to deliver a nuclear weapon, assuming it would have weighed something like 5 tons at the time. While there is no evidence that this was a consideration, it is possible that this prospect concerned Allied military planners. It was not yet time for nuclear missiles to fly in space, beyond the planet's atmosphere, but that time would come, soon enough.

fast aircraft

In the early part of World War I, it became clear that if aircraft fought each other, the faster of two aircraft would be able to out-manoeuvre a slower one, all other things being equal. As a rule, things were never entirely equal, but they were close enough. In any case, having an aeroplane that could turn around quickly was of little use if your quarry could get away from you, so a lot of store was set by having a fast aircraft.

In these days of guided missiles, the dogfights of the World War I variety are no longer seen. Speed is needed for other reasons, mainly to do with getting aircraft (both bombers and fighters) to where they are needed (and away again) in a short period of time. We take speed for granted, but it is speed which has changed the shape of warfare, just as surely as the crossbow did away with knights in armour, and cannons changed the shape of city walls.

Even helicopters, selected because they can hover, need to move fast, or they need to keep moving for a long while. Here are the three most outstanding in that class: two of them are also used for civil purposes

Bell 206 L-1 240 km/h (150 mph)	This helicopter set around-the-world helicopter speed record of 17 days, 6 hours, 14 minutes and 25 seconds in 1996.
Westland Lynx ZB-500 399 km/h (249.09 mph)	Fastest helicopter (world record). This helicopter has achieved Mach 0.3, and has climbed to an impressive 3200 m (2 miles, or 10,600 ft).
Mi-24 Hind 336 km/h (210 mph)	Fastest attack helicopter.

Aircraft have changed our lives in so many ways: here are just a few of the most important society-changers.

long-haul aircraft

In 1903, the Wright Brothers' first flight lasted 12 seconds, too short a distance for accurate timing, but since the plane did not stall, it must have reached 50 km/h (31 mph) while it was off the ground. It covered 37 m (121 ft), so the average was a mere 11 km/h (6.8 mph).

By 1905, their Flyer III was able to cover 39 km (24.2 miles) in 38 minutes and 3 seconds, an average of just over 61 km/h (38 mph).

In 1911, Walter Brookins flew 301 km (187 miles) from Chicago to Springfield, Illinois, making two stops. His flying time was 5 hours and 49 minutes, giving him an average of about 51 km/h (32 mph).

In May 1919, Lieutenant Commander A.C. Read flew an NC-4 flying boat across the Atlantic. It is recorded as having a maximum speed of 145 km/h (91 mph). In July that year, Alcock and Brown flew nonstop in a Vickers-Vimy from Newfoundland to Ireland, 3154 km (1960 miles) in 16 hours, a respectable 200 km/h (125 mph). The aircraft had a maximum speed of around 144.5 km/h (90 mph), but it was helped by a tailwind of

Figure 10. Fast planes

Make	km/h	mph	Qualifying feature
Hawker Hurricane, 1935	506	315	First plane to exceed 300 mph, World War II fighter.
Messerschmitt Bf109, 1935	550	342	World War II fighter.
Spitfire Mk I, 1936	563	350	Battle of Britain fighter.
MC-72	706	439	Fastest single prop driven seaplane.
L-188 Electra	713	445	Fastest initial attack air tanker in service today.
XA2D Skyshark	801	500	The fastest single-engine turbo-prop.
F-8F Bearcat	846	526	Fastest single prop driven aircraft.
Tupolev Tu-114	873	542	Fastest turbo-prop transport.
Ilyushin IL-76	897	560	Fastest operational air tanker.
Tu-95 Bear	921	575	Holds the unofficial record as the fastest multi-engine propeller aircraft.
Boeing 747	969	605	Fastest operating commercial jet.
Tupolev Tu-160	2211	1380	Fastest heavy bomber.
Tupolev Tu-144	2488	1553	Fastest supersonic passenger jet.
FB-111 Aardvark	2964	1850	Fastest US strategic bomber.
F-15 Eagle	3004	1875	Fastest US jet fighter.
MiG 25 Foxbat	3388	2115	The fastest jet fighter in the world.
SR-71Blackbird	3673	2293	The fastest jet aircraft in the world.
X-15	7241	4520	Fastest manned aircraft and also the highest flying one.

about 48.3 kmh (30 mph): a return journey under those conditions would have taken twice as long!

In December 1919, Keith and Ross Smith, flying Vickers-Vimy G-EAOU, reached Darwin, in northern Australia, after a 135-hour long journey, averaging just under 160 km/h (100 mph).

In 1920, a transcontinental air-mail service flew from New York to San Francisco in 33 hours and 20 minutes. That represented 4137 km (2571 miles) at an average speed of 124 km/h (77 mph). In 1923, Oakley Kelly and John MacReady flew from Long Island to San Diego, 4265 km (2650 miles) in 26 hours and 50 minutes, almost 160 km/h (100 mph).

In 1924, it was the right time for a round-the-world flight, and this was achieved by three 2-seat Douglas World Cruisers which completed 44,085 km (27,553 miles) in 175 days, with a flying speed of 120 km/h (74 mph) – over 371 hours and 11 minutes in the air.

Charles Lindbergh's solo flight over the Atlantic in 1927 took 33½ hours at 120 km/h (108 mph). When Bert Hinkler flew solo in his Cirrus Mk II Avian from England to Australia in 1928, he completed 18,427 km (11,450 miles) in 129 flying hours with an average speed of 132 km/h (82.5 mph). A little later that year, Charles Kingsford-Smith flew from San Francisco to Brisbane, Australia, with stops in Hawaii and Fiji at an average speed of 150 km/h (95 mph). Charles Lindbergh set another record with Anne Morrow when they flew from Los Angeles to New York, a distance of 3943 km (2450 miles), in 14 hours and 45 minutes at 265 km/h (166 mph).

Pan American Airways opened an Atlantic service in 1939, flying at a cruising speed of 300 km/h (188 mph). In 1945, a Douglas C-54E Globester carried nine passengers 37,251 km (23,147 miles) around the world from Washington to Washington in 149 hours and 44 minutes (about 250 km/h, or 150 mph).

On 1 December 1947, a Qantas Lockheed Constellation aircraft took off from Sydney. After overnight stops in Singapore and Cairo, combined with four other refuelling stops, it arrived in London, 4 days later, after 55 hours of actual flying time. The distance covered was around 21,726 km (13,500 miles), giving it a flying speed of 389 km/h (245 mph).

By 1952, when the first DeHavilland Comet came into service, its speed of 780 km/h (490 mph) was far greater than that of the Douglas DC-6B which flew at 500 km/h (315 mph). The Comets though had a design flaw that made the aircraft break up from metal fatigue, leaving a gap in the market for Boeing.

The Boeing 707 was tested in 1954 and entered commercial service in 1957. This had a cruising speed of around 960 km/h (600 mph). The Boeing 747 entered service in 1966, with about the same (or slightly slower) speed. This remains the normal speed for a jetliner today as the emphasis has been more on adding extra capacity than on getting anywhere faster. The Concorde entered service in 1971, and before it was withdrawn in 2003, flew from London to Boston in 3 hours, 5 minutes and 34 seconds. The 5250 km (3262 miles) was flown at a staggering 1690 km/h (1055 mph).

By contrast, when Gossamer Albatross became the first human-powered aircraft to fly across the English Channel on 12 June 1979, it made the 35.8 km (22.2 mile) crossing in 2 hours and 49 minutes. It reached a top speed of 29 km/h (18 mph) and flew at an average altitude of 1.5 m (5 ft) above sea level. A spin-off from this, Solar Challenger, flew from Corneille-en-Verin airport, north of Paris to RAF Manston, south of London, 262 km (163 miles) in 5 hours, 23 minutes – an average of 50 km/h (31 mph) – on photovoltaic batteries alone.

And in 1999, Brian Jones and Bertrand Piccard flew their Breitling Orbiter 3 balloon around the world, covering 74,908 km (46,759 miles) in 19 days, 21 hours and 55 minutes, turning in an average speed of 156 km/h

(97.84 mph). We have to ask ourselves – are the modern technologies at a stage equivalent to that of the Wright Brothers, or that of the jetliner, which has hardly changed since the first Boeing 707 flew in 1954?

rocket launchers

The bazooka of World War II was one of the great levellers in land warfare because it allowed infantry the opportunity to tackle tanks, and even cripple or destroy them at times.

Bazookas have the ability to hurl a lump of explosive further than the arm can throw, and they do it from a light and low-tech piece of equipment (the first launcher was made from recycled fire extinguishers!).

The impressive effect of the bazooka came from the huge speeds developed in the shaped charge that was exploded against the target. In the 1880s, an American named Charles Munroe was testing the effects of explosives on steel. By chance, some wads of guncotton had identifying lettering pressed into them. The researchers were amazed to find that this lettering was printed in reverse onto the steel plates that they had just blasted. This was the very beginning of the 'shaped charge'.

A shaped-charge warhead has a suitable explosive formed around a hollow metal liner, usually copper or aluminium. Here, 'suitable' means having a high-speed detonation wave that moves forward, sending a narrow jet of molten, high temperature microscopic particles and plasma at the target. The wave is in the hundreds of kilobars pressure range (a car tyre has a pressure of around two bars), and it moves at about Mach 30. Shaped charges will blast holes in just about anything that they are held against, but how do you 'bell the cat' and hold an explosive charge against a tank fitted with cannon and machine gun? Simple: you sneak up, and send the shaped charge in on a rocket. Speed wins again.

in a spin

Sometimes, things that do no more than spin, rotate, turn or revolve still move quite fast in one place. The classic position of the inadequate bureaucrat in an emergency is a perfect example, as they sit in the corner, emitting a whirring noise. Of course, when gyroscopes spin, it keeps them stable, so perhaps this is what inspires the spavined spinning pen-pusher. It might also explain why so many of them are boring. Trust me: I spent many years living with and studying that tribe, and I know their mores and beliefs as only an anthropologist or a former inmate can.

The first aviation objects to approach the speed of sound, spinning, were the propeller tips on aircraft from the 1920s and 1930s. The planes themselves only reached about Mach 0.5 in the 1930s, but propeller speeds hit this natural limit much earlier – jet engines were needed to take aircraft past the sound barrier.

In the earliest part of the steam age, engines were fitted with governors of the sort invented by James Watt. When an engine went faster these governors responded by reducing the steam flow, while if the engine slowed, the steam valves were opened wider. In this way, a whole culture of engineers acquired the idea of feedback.

In a sense, the need to control the speed of fast-spinning steam engines prepared the way for an information revolution that is still going on – no pun intended.

rotation in the solar system

Everything in the solar system needs to be in orbit, because if a body did not orbit something, gravity would drag it together with something else. Most space bodies also spin, though sometimes this motion is slow and sometimes it is hidden. The Moon rotates so slowly that we always see the same face pointing at us, a phenomenon called captured rotation.

The Moon's structure is ever-so-slightly uneven, and over billions of years, the tug of Earth's gravity on the uneven bits has caused tidal effects that have stabilised the Moon's spin. This means that as the Moon orbits the planet once a month, it rotates exactly once. It may seem to go around once a day: most of its movement across our sky comes from the Earth's rotation.

Captured rotation (or gravitational lock) happens to satellites quite often. The five inner satellites of Jupiter all rotate once for each orbit of Jupiter. So do Deimos and Phobos (moons of Mars), the nine moons around Saturn, and the four moons of Uranus. Triton orbits Neptune in matching time and Pluto's companion, Charon, is in step with Pluto. It even happens to one of the planets, though just a bit differently: Mercury completes three orbits of 88 Earth days around the Sun while rotating twice on its axis.

If a planet is caught up by gravitational lock, the amount of variation on the planet will be greatly reduced. Stability is all very well, but if life like us is to evolve on a planet, you need day and night, seasons, the wetting and drying caused by tides and also the changes caused by weather. All of those are driven by a planet that spins faster than it orbits around its star.

As an example, red dwarf stars have a very narrow zone where life can exist, a band called the habitable zone. Planets in that zone are likely to end up in a 1:1 lock, which will mean they stagnate. Without day and night and seasonal changes, hot spots would become hot and stay hot. On a gravitationally locked planet, water would migrate to the cold side and turn into ice.

Some theories about how life got started on Earth require the operation of tides in a primordial ocean that acted as a sort of biochemical soup, but these are largely guesswork. All the same, most life as we know it is affected by the seasons, day and night and the tides.

Oddly enough, captured rotation, the slowing down of orbiting bodies, is a two-way thing related to the way tides are generated. Just as the Earth's gravitational pull slowed the Moon down to rotate once every 27.32 days, the Moon is slowly working on our planet to produce a tidal friction that is slowing us down. We are just lucky that the Moon is a lot smaller than our planet!

The fastest-spinning object found in the solar system so far is a near-Earth asteroid called 1998 KY26. The tiny asteroid, just 30 m (98.4 ft) across, spins once every 10.7 minutes, which is 10 times faster than any other object and 60 times faster than the average for all of the known asteroids. The unusual speed means that it must be a piece that was split off from a larger asteroid by a collision.

The asteroid is too small to see, but variations in the body's brightness, as observed from several telescopes, confirmed the rotation speed. Radar echoes revealed the asteroid's size, shape and more: it is an almost spherical bare rock with meteorite cratering.

escape velocities

For a planet of a given mass and density, there exists an escape velocity, a speed that a bullet or other object must travel at if it is to escape the surface gravity of that planet forever. Jules Verne had some ideas about this when he wrote *From the Earth to the Moon* (1865), where he proposed sending people to the Moon in a sort of tin can fired from a large gun.

If anybody tried this method, they would have ended up as tinned soup. A long barrel would have made the acceleration more gentle, but even if the barrel could be as long as the Eiffel tower, the acceleration force would still need to be 3500 times as great as Earth's gravity. Making the barrel as high as the Empire State Building would reduce the force to 3000 times that of gravity, usually referred to as g. The most a human can take is about 10 g, so the gun solution would never work.

Verne probably knew that it wouldn't work, but he just needed a transport method to hang his story on. He certainly knew about escape velocity and what the Earth's escape velocity was:

> If the projectile kept indefinitely the initial speed of 12,000 yards a second, it would only take about nine hours to reach its destination; but as that initial velocity will go on decreasing, it will happen, everything calculated upon, that the projectile will take 300,000 seconds, or 83 hours and 20 minutes, to reach the point where the terrestrial and lunar gravitations are equal, and from that point it will fall upon the moon in 50,000 seconds, or 13 hours, 53 minutes, and 20 seconds. It must, therefore, be hurled 97 hours, 13 minutes, and 20 seconds before the arrival of the moon at the point aimed at.

Verne was also clever enough to see that there was an advantage in having a launch site near the equator, and placed his fictional gun in Florida, where Cape Canaveral, the Air Force station and NASA space centre, is found today. At 28° north, this is not as good as the 4° north of the European Space Agency's Kourou site in French Guiana, but it beats the Russian Baikonur Cosmodrome, which is at 47° north in Kazakhstan. An equatorial launch means the rocket already has a speed to the east of about 1650 km/h (more than 1000 mph), with respect to the Earth's centre, before the engines start. That means that the rocket needs to burn less fuel to get into orbit, or to reach escape velocity.

It is not only planets that have escape velocities: if a spaceship is ever to escape our galaxy, it needs to build up a speed of about 1000 km/s (621 miles per second), 3.6 million km/h (2.24 million mph) to get out of the Milky Way. Escaping our planet is about one-ninetieth of that at 11.2 km/s (6.95 miles per second): our planet has much less mass, but we are much closer to the mass, and distance counts.

the statistics of the solar planets

Escape velocity depends on a number of things: a less dense planet will have an equator that is further away from the centre, reducing the escape velocity. A more massive planet of the same density will have a stronger gravitational pull, which means a greater escape velocity, and so on. Jupiter is far less dense than most other planets but far larger, so its huge mass still overcomes those other factors (see Figure 11 on page 139).

fast extrasolar planets

Just a few years ago, the very idea of spotting planets around other stars would have seemed like the science of Cloud Cuckoo Land. Even with a telescope like Hubble, safely out in space and free of the interference caused by air currents, no telescope in existence could 'see' a planet near another star. That remains true today for most planets, but what has changed is how people can use other methods to 'see'.

Imagine two skaters of similar height and weight with their hands clasped, swinging around on the ice. You don't see one circling the other,

Figure 11. Escape velocities

To escape:	km/s	km/h	mph
The Milky Way	1000	3,600,000	2,248,595
Earth's equator	11.2	40,320	25,184
Jupiter's equator	59.5	214,200	133,791
Mars' equator	5.0	18,000	11,243
Mercury's equator	4.4	15,840	9894
Moon's equator	2.4	8640	5397
Neptune's equator	23.6	84,600	52,842
Saturn's equator	35.5	127,800	79,825
Uranus' equator	21.3	76,680	47,895
Venus' equator	10.4	37,440	23,385
Sun's equator	617.5	2,223,000	1,388,507

you see them circling a central point, where their hands meet: they both orbit around their common centre of gravity. If you replace one skater with a Sumo wrestler, the rotation point moves. The wrestler would still move a bit, but it would look like the other skater was almost in orbit around him. With a lighter second skater, the effect would be enhanced.

When a planet orbits a star, they both rotate around a central point, and the star seems to wobble a little like the Sumo wrestler would. When the planet is moving away from us, the star is coming our way, and vice versa. That means we can look for Doppler shifts that indicate the presence of a planet, and can learn its orbital period and distance.

Just 15 light years away, a rocky world, 7.5 Earth masses and about twice the Earth's diameter, whizzes around Gliese 876 (or GJ 876), a star just like our Sun. It does so a mere 3 million km (just under

2 million miles) away from the star, which means it is rather hotter than our planet, to say the least of it. You could bake a roast on it.

Our sort of life would be impossible, with years taking just 1.94 of our Earth days. Then again, maybe life in the fast (p)lane(t) is different! There may also be other smaller and slower planets still to be found around GJ 876.

black holes and the galactic zoo

It is hard to understand the very small, the world where quantum effects happen, but it is equally hard to understand the very large, the very fast or the very heavy, where black holes and relativity take control. It is easier perhaps to just look calmly at the physicists and astronomers and nod our agreement.

By rights, given what we know now, the Milky Way galaxy ought to fly apart as it spins too fast. It doesn't, so somewhere in the centre of the galaxy, there must be a HUGE black hole, a monster collection of matter, millions of times more massive than our Sun, that is slowly gobbling more and more material.

In 1998, Farhad Yusef-Zadeh reported a peculiar stream of gas, close to the centre of the galaxy. This plasma stream was travelling at a remarkable speed. Because of its speed, it appears to be moving close to a very massive object, but then it swings around and flies off on a new line, like a runner swinging around a steel pole in the ground.

At its highest speed, Yusef-Zadeh said, the gas travels at more than 3 million km/h (under 2 million mph). While impressive, it is less than one-three-hundredth of the speed of light, barely fast enough for relativity to come into the picture. All the same, if the plasma ran into anything else, the interactions would be highly interesting. And deadly to nearby life.

Figure 12. Planetary data

Planet	Radius (Earth = 1)	Mean distance from Sun (AU)	One solar orbit in Earth years	Orbit speed in km/s	Mass (Earth = 1)	Density (water = 1)	Escape velocity in km/s
Mercury	0.38	0.386	0.241	47.9	0.0558	5.6	4.4
Venus	0.95	0.72	0.615	35	0.815	5.2	10.4
Earth	1	1	1	29.8	1	5.52	11.2
Mars	0.53	1.52	1.88	24.1	0.107	3.95	5
Jupiter	11.21	5.187	11.9	13.1	318	1.31	59.5
Saturn	9.46	9.533	29.5	9.64	95.1	0.704	35.5
Uranus	4.01	19.133	84	6.81	14.5	1.21	21.3
Neptune	3.88	30	165	5.43	17.2	1.67	23.6
Pluto	0.18	39.333	248	4.74	??	??	0.9?

Pluto stays in the table, even if the IAA have downgraded its status. Pluto is still interesting – the author is a member of 'When I was Your Age, Pluto was a Planet' and who knows, it may get promoted again one day! Some columns include data from other pages (such as escape velocity) to provide a single summary table of the main information associated with the planets.

Luckily, the whole thing is 25,000 light years away. Even if the gas started to head straight at us without being slowed down by gravity, it would still take 9 million years to reach us. We should worry more about being sucked into a black hole, as the theory is that the tidal forces will stretch you out thinner than spaghetti. Just keep looking calmly at the page, nodding in agreement ...

Yusuf Zadeh is fond of saying that the centre contains a galactic zoo, of which the gas stream is just one curiosity. In 1987, he and John Bally made another interesting discovery in the centre of the galaxy. While astronomers call this conical-shaped object G359.3-0.82, it is slightly better known in amateur and science fiction circles by Yusuf Zadeh and Bally's name for it: 'the Mouse'.

Consider what happens when a powerboat travels fast over the water. It leaves behind a V-shaped wash (or wake) that has an angle determined by two things: the speed of the boat and the speed of the water waves. As the wave speed is fixed, a satellite image of the angle of the wash would let a space observer calculate the boat's precise speed.

Similarly, when a supersonic aircraft goes by, it leaves a pressure cone behind it. Once again, the shape of the cone is determined by the speed of sound in air and the speed of the aircraft. The wake and the sonic boom are both shock waves.

In 1989, John G. Cramer made a playful but interesting suggestion in *Analog*, regarded by many as the premier science fiction magazine. Suppose, wrote Cramer, that this conical object was a high-speed spacecraft, ploughing through the thin, but hot, plasma at the galactic centre.

He estimated that the speed of sound in the plasma would be about 100 km/s (60 miles per second), and that the angle on the cone indicated that the Mouse was travelling 5.8 times that speed, or 580 km/s (360 miles per second), a bit over 2 million km/h (about 1.25 million mph).

Now comes the worry: Bally and Yusuf Zadeh assumed that the Mouse is some 27,000 light years away, which is reasonable, but not guaranteed. The Mouse might in fact be much closer. The good news is that, for now, that particular bit of the galactic zoo doesn't seem to be speeding our way.

centrifuges

Australians of a certain age still know how to make tea in a billy, a large tin can with a handle. The billy is filled with water and placed on a fire until the water is lumping and bumping nicely, then a handful of tea leaves are thrown in, and the water is allowed a few seconds of extra turmoil before the billy is lifted off the fire and allowed to stand and settle.

A few tea leaves always stay on the surface, and now comes the skilled part. The tea maker has to pick up the billy by the handle and swing it several times in a complete circle, relying on this 'centrifugal force' to stop the water cascading down on the billy-swinger. The last challenge is to stop the billy without losing any tea.

There is no such thing as a 'centrifugal force' in tea making, but it is a polite way of describing a complex situation, and even if the force is a total fiction, it still works as an explanation. The end result of the swinging is that the floating tea leaves, subjected to what is in effect a greater than usual gravitational force, sink to the bottom, out of the way.

As a young chemistry student, I learned to use a centrifuge, a simple hand-cranked device that whirled a test tube around. As the spinning portion moved faster, the tube swung in a pivot until it was almost horizontal. We were taught to add a second test tube with an equal amount of water in it on the other side to keep the whole thing balanced, but that was really just to get us trained for when we used electrical centrifuges.

The idea of the chemical centrifuge is to persuade insoluble precipitates to sink to the bottom when the particles have almost the same density as water and so have little inclination to sink. A hand-cranked centrifuge probably rotates 5 or 6 times a second, around 300 to 400 rpm, while an electrically powered centrifuge might spin at 3000 to 5000 rpm; we also have the ultracentrifuge, spinning at 40,000 to 100,000 rpm.

142

If we assume the centrifuge has a diameter of just 10 cm (4 inches), that means that each spin of the ultracentrifuge sees the load travel 31.4 cm, just over a foot, meaning the load travels between 13 and 40 km (8 and 25 miles) a minute, between subsonic and supersonic.

A laboratory centrifuge is rated not by revolutions per minute, but by its relative centrifugal force (RCF), in Earth gravity (*g*) units by inserting the radius (r) and the RPM into this equation:

$$RCF = 1.1118 \times 10^{-5} \times r \times RPM^2$$

Large slabs of modern biochemistry, biotechnology and genetics hang from the use of centrifuges, while larger and slower centrifuges have been used in the past to test the resistance of astronauts and cosmonauts to *g*-forces. A modern commercial centrifuge will deliver about 30,000 *g*, but humans can only take about 10 *g* before they black out. We are bags of liquid, and *g*-forces stop blood from finding its way back through veins to the heart.

In the future, centrifuges might just have a different role in space. It may also never happen, but in 1998, Clifford Singer and his colleagues published one such idea in *Acta Astronautica* outlining a way of using orbiting centrifuges to throw ball-bearings. They said that the forces might be used to boost satellites into higher orbits, launch spacecraft to distant planets, or slow satellites and returning space probes for safe re-entry into the atmosphere.

The idea was to use 'centrifugal relays' to fire re-usable ball-bearings back and forth, providing thrust without needing bulky rocket engines and huge loads of fuel – and avoiding the release of foul and toxic rocket exhaust gases into the atmosphere or near-Earth space. A primary centrifuge or chain of centrifuges would accelerate the ball-bearings, which would be 'caught' by a secondary centrifuge mounted on the spacecraft, transferring their momentum to the payload.

143

The ball-bearings would then be hurtled back at the primary centrifuge, both to be re-used and also to give the spacecraft extra momentum (remember dear old Newton's equal and opposite reaction?). They reported that the projectiles could be accurately aimed and tracked over 10 km (6.2 miles).

Well, we must wait and see, but there are more practical uses. A monster centrifuge at Boulder, Colorado can swing 2 tons at 200 g, and this has been used to test scale models of dams. But what about paint stripping by centrifuge? Would you believe that? At Oak Ridge, Tennessee, engineers have used a centrifuge to throw tiny pellets of frozen carbon dioxide at the speed of sound to strip paint and contaminants from surfaces, without solvents or abrasives that need cleaning up.

Oak Ridge, of course, is mainly a centre for nuclear technology, and that reminds us that centrifuges once changed history. The uranium needed for weapons must be enriched, so as to increase the relative amount of uranium-235 (U-235). The easiest way to achieve this is to make uranium into a gas, uranium hexafluoride, and then to separate the U-235 variety from the more common U-238 variety. In 1944 and 1945, this method was used to prepare the bomb that was dropped on Hiroshima, and when you hear of 'rogue states' trying to enrich uranium today, they are generally also using centrifuges.

the patented centrifugal birth table

Administrators look down on scientists as unworldly, scientists look down on engineers as uncultured, engineers regard accountants as impractical, accountants look down on actuaries as impersonal, and actuaries, the only ones with sense, look down on administrators.

As a scientist, I'd observe that only a childless engineer could possibly come up with the idea of a centrifugal birth table. This is not only my conclusion, but the view of those of my peers who gather annually to award the Ig Nobel Prizes for the sort of research that makes people both laugh and think.

To have come up with this idea, George and Charlotte Blonsky must have been childless. George was an engineer, and saw the world through an engineer's eyes. When he saw a pregnant elephant at the Bronx Zoo, he noted that she was twirling herself around. A baby elephant is about 120 kg (264 lb), more than most adult humans, and therefore something of a challenge to deliver, even for an elephant. George was inspired.

Women have trouble giving birth, clearly, and the problem was the application of forces. He could see a straightforward engineering solution. Strap the mother to a table, whirl it around, and let physics do the rest. He and Charlotte drew up specifications, plans and blueprints, and in 1965, gained US patent 3,216,423 for their birthing table.

Giraffes drop their babies onto the veldt from 2 m (6.5 ft), giving them a velocity of 20 km/h (12.5 mph) on landing, and baby giraffes survive. Why do I have an image of a whirling table surrounded by a ring of nervous people kitted out with catching gloves? Silly me, George was no fool: he included a safety net!

making reactions go faster

There are several secrets to understanding chemistry, but chief among them is the idea of equilibrium. Atoms and molecules whirl around, they join up and disconnect, they change partners, but in the end, a balance is always achieved, where the changes going one way are balanced by changes going the other way.

Industrial chemistry relies on either changing those balances by applying heat or pressure, or by extracting a desired product as it is formed. At times, though, the most important goal is making the system reach its equilibrium point sooner. If you put nitrogen and hydrogen in a container under fierce pressure, over time, the molecules will respond by reconnecting to form ammonia.

The problem is that it will take a perilously long time for this to happen, far too long for the operation to be profitable. On the other hand, if you expose the gases to expensive osmium, or cheap porous iron, made by reducing magnetite, you provide sites where ammonia can form really easily, and the process reaches equilibrium very fast. The final yield is the same, but you get there sooner.

When Fritz Haber developed this process in 1909, he had no intention of aiding a war effort, any more than the people who developed the centrifuge thought that they might be providing the tools to make horrible weapons (or George Blonsky's birth table). Haber wanted to benefit humanity, and help find a solution to the need for 'fixed nitrogen' in agriculture, compounds that could be used to make fertilisers.

You could start with saltpetre, an impure sodium nitrate that was mined in Chile, South America, and which could be loaded into lumbering sailing ships that hauled it around Cape Horn and up the Atlantic, but that cost time and money for labour.

The solution was to find a way of synthesising nitrogen compounds, and that meant finding a way of speeding up the time it would take for the hydrogen, nitrogen and ammonia to reach equilibrium. The catch was that when WWI started, Britain was fairly well equipped to control the seas, and could certainly have stopped any Chilean saltpetre reaching Germany – the saltpetre had one other main use, as a feedstock for explosives.

Haber's little catalyst allowed Germany to stay in the war far longer than they might have done otherwise. Without it, the Russian revolution might not have been so extreme, Germany's defeat would have been less galling and Mr Hitler might not have found it quite so easy to grab power. We will never know for sure.

Catalysts make all sorts of other reactions go faster as well. Living things use enzymes which act as catalysts to make things happen inside the cell that would never happen in the time available. The definition of a catalyst is that it changes the speed of a reaction (strictly, the time taken to reach equilibrium) without being changed itself. Our lives rely on enzyme catalysts, both to keep us going and also to grow the things we eat. Admittedly, some of the uses are a bit more frivolous.

In 1824, rich English gentlemen were alerted to the availability of a clever little cigar lighter in an era before there were any reliable matches. A small bottle contained acid and zinc metal, which were allowed to join to form hydrogen, which was then allowed to stream across a fine platinum wire. When hydrogen and oxygen come into contact with finely divided platinum, the hydrogen reacts quickly with the oxygen to form water and heat. Döbereiner's Lamp creates enough heat to make the platinum glow red-hot, which causes the hydrogen-oxygen mix to burst into flame, providing a quick way to light a gentleman's cigar.

the slowest biological reaction

Every biochemical reaction in our bodies is managed by an enzyme, or often, a chain of enzymes. These work as catalysts to make reactions happen faster than they would otherwise do; the scale of the acceleration is truly staggering. One particular reaction, the breakdown of phosphate monoesters, would take a trillion years if it had to happen all by itself, rather than the microseconds or milliseconds it takes now.

Richard Wolfenden had already found a biochemical change that is absolutely essential and which, if left to its own devices would take 78 million years in water. He has since found a key reaction so slow that if you put the ingredients in a test tube without a catalyst, only half of them would be used up after a trillion years, about 200 times the age of the Earth.

All living cells function as they do because of signalling that goes on in the cell, and phosphatase enzymes are a part of this. Some of the enzymes break down chemicals called phosphate monoesters, while others help in the extraction of smaller carbohydrates from starch or in sending hormone signals. It would be a long time between signals of wellness or feeding or whatever if the enzymes did not do their job. In fact, Wolfenden says there is a difference of 21 orders of magnitude. That is, the catalysed reaction is 10^{21} times as fast.

According to Wolfenden, it shows just what natural selection can achieve over the millennia, and it sets a few benchmarks and targets for people working in drug development. When you have a catalyst that kicks reactions along, it is an excellent place to start development work. In the next few decades, a lot more should come out of people tailoring enzymes and tinkering with enzymes. I decided to call this study enzymomics, but I am too late: there are a few web references to the term already. Curses, foiled again!

a speedy death

When the Bulgarian dissident Gyorgyi Markov was shot with an 'umbrella gun' in London in 1978, the aim of the shot was to lodge a small pellet containing a catalyst protein called ricin beneath his skin.

Ricin is a particularly nasty poison that is derived from the castor bean. It is popular with journalists seeking to do beat-ups about terrorism, because it is remarkably deadly, and it is easy to obtain. Luckily, the details about how to prepare it are a little harder to come by, and ricin is more likely to save lives than to take them. It is a neat little catalyst that takes a cell and undoes its insides, a bit like a tiny saboteur running around inside some electronics, snipping and snapping bits off, here and there. In actual fact though, ricin is far more disciplined than that.

Ricin has two parts, two chains that we will call A and B. The A chain is globular and 267 amino acids long. The B chain is shaped like a barbell and 262 amino acids long. The B chain hooks onto a cell, and triggers a reaction by the cell that takes the whole molecule into a bubble called a vacuole, inside the cell, where the chains separate. The B chain opens a gap to let the A chain loose, and it proceeds to take apart the cell's ribosomes, the parts that make proteins. The A chain is a catalyst, and as it finishes off each ribosome, it lets go and attacks another.

And how might this save lives? The secret is that cancer cells have marker molecules on the outside. A number of researchers are looking at making magic bullets out of a ricin molecule and a special biochemical tag that will attach only to the cancerous cells and then let a single ricin molecule loose. The theory is sound, but the practical side of terrorising cancer cells is still to come.

sensing rotation

Our senses are remarkable things, but probably nothing is more remarkable than the way we detect rotation, using the three tubes, the semicircular canals, that form part of each of our ears. Even with our eyes shut, we can tell when we are turning. But, if we spin around fast, we seem to be still turning when we stop. The answer lies in the operation of the semicircular canals.

The three tubes lie more or less (but not exactly) at right angles to each other, and each contains fluid and remarkably sensitive little bristles. If you don't appreciate how sensitive hair cells can be, close your eyes and get somebody to touch one hair on your arm with a toothpick. Not the arm itself, just the hair.

Do you know how to tell a hardboiled egg from a raw one without opening it? The trick is to spin the egg on a plate, then stop it quickly and let go. In a raw egg, the inside is still spinning, and when you let the egg go, it will mysteriously start spinning again, while the solid insides of a hardboiled egg don't do this. (Note: The plate is to stop spun eggs falling on the floor!)

When you spin around, the fluid in the canals stays still, so the hairs detect motion, and provide a signal to the brain, with the force giving an indication of how fast you are spinning. If you keep turning, the fluid starts to spin as well, and when you stop, the liquid keeps moving, providing a false signal that you are still moving – giddiness comes from your brain trying to resolve this conflict.

One thing though: a boat can roll, pitch or yaw, involving motion in three different spatial dimensions. That may give you a hint as to why we have three canals. And when we are at sea, rolling, pitching and yawing, the turmoil in our semicircular canals confuses our brains and makes us seasick. Watch the horizon!

the speed of a nerve impulse

Most school-level science texts explain nerve impulses as a sort of electrical signal. It's not a bad analogy, but it isn't a good one either. It explains why, if a nerve is cut, a signal does not go through, but it leaves out a lot of delicious detail about a neat method of chemical signalling, and a great deal about speeds.

The simplest chemical signal involves one cell making a chemical, which diffuses to the next cell and so on. This is slow: if a *T. rex* attacked the tail of a *Brontosaurus* (OK, *Apatosaurus*, if you insist), it would take 30 years for Bronto's brain to find out by diffusion what was happening to his tail.

Nerves are hardwired connections joining two places. The brain is a network (often compared in schoolrooms to a telephone exchange) where new connections can be made if necessary. What a nerve really does is send a wave of reaction, with potassium ions moving out of the nerve cell and sodium ions moving in. This produces something that is in many ways like an electrical signal, except that it isn't. The start is a chemical message reaching one end of the nerve, the end is a chemical leaving the other end – and it is a one-way system.

Pain signals travel at about 0.6 m/s (2 ft/s). Some nerve signals that trigger muscular movement can go at over 100 m/s (328 ft/s), while sensory impulses hit 30 m/s (98 ft/s), the difference being the myelin sheaths wrapped around the nerve fibre. In schoolroom-speak, the myelin works like insulation on an electric cable, and that analogy sort of works, but only sort of!

Now back to our *Brontosaurus*: at about 90 m (295.3 ft), a big Bronto will not feel the pain of *T. rex*'s bite for up to 150 seconds. Maybe the big ones learned to keep an eye on their tails, or maybe they just had better myelin sheaths than we do.

reaction time

How long does it take you to react? A simple method of testing this is to have somebody drop a metre stick while your finger and thumb are on either side of the stick. As soon as you see the stick move, you grab

Figure 13. Reaction times

Distance in centimetres	Time in milliseconds	Distance in centimetres	Time in milliseconds
10	0.143	30	0.247
20	0.202	32	0.256
22	0.212	34	0.263
24	0.221	36	0.271
26	0.230	40	0.286
28	0.239	50	0.319

it, and then see how far the stick has gone past your finger and thumb. You can convert this distance to a time where s is the distance, a is the acceleration due to gravity (9.8 ms^{-2}, or 32 ft/sec/sec) and t is the time in seconds. The formula is s = 1/2at^2, but to make life easier, see Figure 13, based on t = $\sqrt{(2s/a)}$.

The catch here is that you might be tempted to try and guess when the person dropping the stick will let go. You will need to take about 20 measurements and ignore the highest and lowest two or three cases. The rest should cluster together.

To try to assess how long it will take you to react when you need to brake a car, you probably need to hold your hands about 20 cm (8 inches) either side of the metre stick, but then measurement gets a bit harder. Now there's a challenge for you!

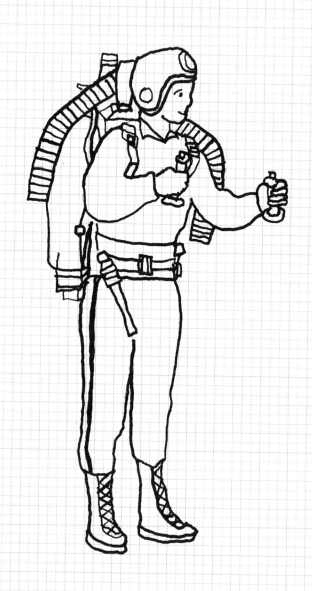

'what goes up ...'

In this chapter, I ended up worrying that I was pulling too many rabbits out of a very worn hat, so I decided to go back and show where the rabbits come from. I've tried to keep the mathematics light enough for readers to follow, but if you prefer, you can ignore all of this and take it on faith.

Still with me? OK, the secret is Favmust: the seven values that we need to plug into various equations to get other values. To keep things simple, I will use metric (SI) units only:

F is the force in newtons;

a is the acceleration in metres per second per second (m/s/s or ms^{-2}): at the Earth's surface, the acceleration due to gravity is 9.8 ms^{-2} ;

v is the velocity in metres per second (m/s or ms^{-1}) at the end of time t;

m is the mass in kilograms;

u is the starting velocity (why? Dunno – it's tradition!);

s is the distance in metres (again, it's tradition!);

t is the time in seconds

Those variables are linked in a series of equations, all of which can be fiddled, so that $F = ma$ (this also means $a = F/m$ and $m = F/a$). Then we have: $s = ut + 1/2at^2$, $v^2 = u^2 + 2as$ and $v = u + at$.

If something starts out not moving ('from rest' in physics-speak), then the value of (u) is zero, so the speed of a freely falling object is found by multiplying the acceleration (a) by the time in seconds (t). As to how we got those equations, that remains as too-hard rabbit-out-of-hat stuff!

it all started with galileo

Galileo Galilei wanted to study how things fell and accelerated, but in the early 1600s, timing was the problem. The watches and clocks were too inaccurate. Easy to fix: Galileo started rolling and sliding things down a slope. It was like having his very own personal slow motion system.

Galileo knew instinctively that if something fell through an altitude drop of 2 m (6.5 ft), whether it went sideways down a slope, or plummeted straight down, it would have the same speed after dropping the same distance. Of course, this was a bit wrong: friction would slow the weight sliding down the slope, but he could get an idea of sorts about how things accelerated under the pull of gravity.

Forget the popular yarn that Isaac Newton 'discovered gravity' when he saw an apple fall: people knew about gravity before that. All Newton did was to work out the law which described how gravity worked; we will come to that in 'Isaac's apple', see page 161.

Galileo found that as things fall, they accelerate uniformly. A dropped weight is travelling at 9.8 m/s (32 ft/s) after 1 second, 19.6 m/s (64 ft/s) after 2 seconds, and so on. The rate of change stays the same.

But you know the story about Galileo dropping two balls from the Leaning Tower of Pisa? It never happened. He worked out, without doing it, that they would fall at the same speed, and mentioned in passing that somebody had tried dropping weights from an unnamed tower, adding that they needn't have bothered. That was all a biographer needed to make up a good yarn about Galileo on the Leaning Tower of Pisa. What Galileo *really* did is in the quote that follows shortly.

physics and the leaning tower of pisa

Galileo's reasoning went along these lines, even if he didn't put it in these terms. Everybody believed that heavier objects fell faster than light objects, because a rock fell faster than a feather. But if that was the case, if you dropped off a cliff or a tower, your body would weigh more than your head, so your neck should stretch as your body accelerates away!

Galileo never said it in as many words, but he could obviously feel that air got in the way. After all, Leonardo da Vinci had invented a parachute, more than a century earlier, so the idea of air resistance was in the air, so to speak. Leonardo's parachute though wasn't in the air, because there was nothing to drop it from. A skydiver called Adrian Nicholas did test Leonardo's design in June 2000, jumping from a hot-air balloon at 3000 m (9842 ft).

> If a man is provided with a length of gummed linen cloth with a length of 12 yards on each side and 12 yards high, he can jump from any great height whatsoever without any injury.
> – Leonardo da Vinci, about 1485

Adrian Nicholas said it was a smoother ride than from a modern parachute, but at 90 kg (198 lb), landing under it would not have been much fun, which explains why he cut himself free after five minutes, and landed with a conventional parachute.

The biographer who said Galileo had dropped balls from the Leaning Tower of Pisa was probably thinking of this comment by a fictional speaker created by Galileo:

> But I, Simplicio, who have made the test can assure you that a cannon ball weighing one or two hundred pounds, or even more, will not reach the ground by as much as a span ahead

of a musket ball weighing half a pound, provided both are dropped from a height of 200 cubits.

— Galileo Galilei (1564–1642), *Dialogues Concerning Two New Sciences*, *First Day*, Dover, 1954, p. 62.

The question remains: would the choice of high or low side matter when you were dropping balls from the Pisa tower? We don't have reliable data on the amount of lean that the tower had back then, but the low side of the tower is now 55.86 m (183.27 ft) above the ground, while the high side is 56.7 m (186 ft) above the ground. That's about 60 cubits.

Dropping a real ball from the high side would result in the ball hitting against the side of the tower. But what would happen if we released lead or iron balls from the same *heights* as the high and low sides?

A little bit of simple physics reveals fall times of 3.376 and 3.402 seconds for the low and high sides respectively, ignoring air resistance, a difference of about one-fortieth of a second, too close for us to judge by sound which one hit first, though the separation in distance when the leading ball hit would be 0.84 m (2.76 ft), the same as their original difference in height at the moment they were dropped. We would be able to *see* the difference, but we probably could not *hear* it.

terminal velocity

You can drop a mouse down a thousand-foot mine shaft; and, on arriving at the bottom, it gets a slight shock and walks away, provided that the ground is fairly soft. A rat is killed, a man is broken, a horse splashes.

— J. B. S. Haldane (1892–1964), 'On Being the Right Size', from *Possible Worlds*.

Figure 14. Non-terminal velocities (fall speeds, ignoring air resistance)

Floors	Metres	ms^{-1}	km/h	mph	Seconds
1	3	7.7	27.6	17.2	0.8
2	6	10.8	39.0	24.4	1.1
3	9	13.3	47.8	29.9	1.4
4	12	15.3	55.2	34.5	1.6
5	15	17.1	61.7	38.6	1.7
10	30	24.2	87.3	54.5	2.5
15	45	29.7	106.9	66.8	3.0
20	60	34.3	123.5	77.1	3.5
30	90	42.0	151.2	94.5	4.3
40	120	48.5	174.6	109.1	4.9
50	150	54.2	195.2	122.0	5.5
60	180	59.4	213.8	133.6	6.1
70	210	64.2	231.0	144.3	6.5
80	240	68.6	246.9	154.3	7.0
90	270	72.7	261.9	163.6	7.4
100	300	76.7	276.1	172.5	7.8
110	330	80.4	289.5	180.9	8.2
120	360	84.0	302.4	189.0	8.6
130	390	87.4	314.7	196.7	8.9
140	420	90.7	326.6	204.1	9.3
150	450	93.9	338.1	211.3	9.6

It isn't the fall that kills you, it is the sudden deceleration when you stop. If you are human-sized, a drop of more than five floors will usually either be fatal or leave you badly paralysed. There is just one saving point: even minus a parachute, a falling human has a terminal velocity.

Skydivers rely on this when they do manoeuvres around each other, tucking themselves up to increase their terminal velocity, or spreading their arms and legs to slow their maximum rate of fall.

Terminal velocity is reached when the force downward, caused by gravity, is equal to the upward force of the air pressure beneath you. It is inexact, but an average terminal velocity for something the size of a human is around 300 km/h (180 mph), but just $2/3$ of that speed when the body is spread-eagled.

Even at the highest terminal velocity, three people are known to have survived falls from aircraft minus a parachute: a Russian pilot, I. M. Chisov in 1942, an RAF tail gunner, Nick Alkemade in 1944, and a Yugoslav flight attendant, Vesna Vulovic, in 1972. Alkemade's fall was slowed by trees before he landed in snow, the others landed in snow – a combination of terminal velocity and a slower deceleration saved them.

leaping from a tall building

One of the problems which taxed me as a schoolboy was: 'If you jumped from the 102-floor Empire State Building, would you penetrate the roadway or splatter?' Lacking sufficient data, my friends and I concluded that you would splatter as you went through (well, we *were* schoolboys and physics students!).

To come up with the 'real' answer, we can use $v^2 = 2as$ here, because the initial velocity is zero, while fiddling $v = at$ to $t = v/a$ gives us the fall time, if we ignore air resistance.

There is a principle of science which says that energy can neither be created nor destroyed. Among other things, that means that speed up and down is beautifully symmetrical. If Superman wants to leap over a tall building, he needs to leave the ground with the same speed that he would have if he dropped from the top.

some famous high towers

Assuming that you could get there and get a clean jump, how far would you fall, how fast would you be going and how long would it take (as always, ignoring air resistance) to fall all the way down from a few well-known landmarks?

To forestall quibbles, as you dive in to Figure 15, there is no nose on the Sphinx, though the legend that Napoleon's French soldiers shot it off does not stand up, as a book published by Frédéric Louis Norden in 1755 shows the nose missing before Napoleon was born. And Big Ben is a bell, not the clock which includes the bell.

isaac's apple

The standard view of Newton and his gravitational apple has him sitting in an orchard, seeing an apple fall, and consequently realising that something attracts things to fall towards the Earth. The reality was a bit different, because by that point, everybody already knew that things fall towards the Earth.

Even the ancient Greeks were aware of things falling under gravity, and once they realised, around 500 BC, that the Earth was a sphere, they went a step further. They saw that wherever you were, things fell

Figure 15. Famous falls (ignoring air resistance)

Tower, building or monument, location	Height in metres	Height in feet	m/s at base	km/h at base	ft/s at base	mph at base	seconds of fall
Nose of the Sphinx, near Cairo	20	65	19.8	71.3	64.5	44.0	2.0
Big Ben's clock face, London	55	180	32.8	118.2	107.3	73.2	3.4
Leaning Tower of Pisa low side	55.9	183	33.1	119.1	108.3	73.8	3.4
Leaning Tower of Pisa high side	56.7	186	33.3	120.0	109.1	74.4	3.4
Top of sail, Sydney Opera House	65	213	35.7	128.5	116.8	79.6	3.6
Top of Big Ben's tower, London	96	314	43.4	156.2	141.8	96.7	4.4
Eiffel tower antenna, Paris	324	1063	79.7	286.9	260.8	177.8	8.1
Fernsehturm, Berlin	368	1206	84.9	305.7	277.8	189.4	8.7
Empire State Building antenna, New York	449	1472	93.8	337.7	306.9	209.3	9.6
Petronas Towers antenna, Kuala Lumpur	452	1483	94.1	338.8	308.1	210.1	9.6

straight down, which meant that things always fell towards the centre of the Earth. That was a very good observation, but unfortunately, they also decided that if things fell towards the centre of the planet, the centre of the planet must be the centre of the universe, which meant that everything, including the Sun and the Moon, went around the Earth.

By the time Isaac sat down in his orchard, that sort of model had been tossed out, and scientists had a fair picture of how the solar system worked – or at least how the easier-to-see parts of it worked. Galileo had found four moons going around Jupiter, Kepler had found out that the planets travelled in elliptical orbits, we understood the phases of the Moon and how it orbited the Earth and all was sweet. We knew what was what, even if we didn't know why.

Whether or not Isaac Newton ever did see an apple fall (though I suspect he did), it was still Newton who came up with the crucial 'why not' question: why would an apple fall, when it was quite obvious that the Moon did not fall?

What he saw, almost in a flash, was that the force of gravity must get less as objects become more distant, and he worked out a law that described this. We call it the Inverse Square Law, because that is a mathematical way of saying the force between two bodies is inversely proportional to the square of the distance between their centres of gravity. Newton would have written it as:

$$F \propto \frac{m_1 m_2}{R^2}$$

where (m_1) and (m_2) are the masses, and (R) is the distance between their centres of gravity.

In *Angels and Demons* (2004), Dan Brown has a character assert that gravity at 18 km (about 60,000 ft) is 30% less than at the Earth's surface, which is something of an exaggeration. If we calculate the drop in gravity

Figure 16. The Earth's gravity with altitude

Height from surface (km)	Gravitational pull	% of surface gravity
0	9.8	100%
18	9.772385723	99.718%
50	9.723676012	99.221%
100	9.648531685	98.454%
200	9.501674277	96.956%
500	9.086754003	92.722%
1000	8.470284939	86.431%
5000	5.490413369	56.025%
10,000	3.813439218	38.913%
50,000	1.107433032	11.300%
100,000	0.586875999	5.989%
385,000 (Moon's orbit)	0.159506349	1.628%
150,000,000 (Sun)	0.000416156	0.004%

with altitude, when we are at the surface of the planet, we are about 6370 km (3,958 miles), on average, from the centre of the planet, so an extra 18 km (11 miles) makes little difference at all! Figure 16 shows what happens with altitude:

If you look at that, Dan Brown's figure was out by two orders of magnitude, 30% against less than 0.3%. Still, when you consider some of the other 'physics' in pulp fiction and pulp movies, perhaps that wasn't so bad!

playing golf on 433 eros

As we have seen, golf has been played in space, both on the Moon and from a makeshift tee on the International Space Station. In each case, it was just a single drive, with no attempt to complete the hole, but it occurred to me that it would be interesting to consider what prospect there would be of playing golf on the asteroid 433 Eros.

I can even date this pondering with some precision: it came a few days after 10 January 1999, which was when the NEAR Shoemaker spacecraft passed close to the asteroid, taking photos of the surface. They were fuzzy, but the dimensions were set down: it was now officially 33 x 13 x 13 km (21 x 8 x 8 miles), as opposed to a previous estimate of 40.5 x 14.5 x 14 km (25.3 x 9 x 8 miles). We knew that the asteroid rotates once every 5.27 hours and has a specific gravity (density) of 2.7. Note the long and short dimensions: 433 Eros is built like a can of drink, with a long and a short dimension, and that will become important.

There was a ridge extending 20 km (12 miles) along the asteroid, and this, and the high specific gravity, suggested that Eros is a homogeneous body, not a 'collection of rubble' as some had suspected. More to the point, the surface of Eros has many craters. The two largest craters are 8.5 and 6.5 km (5.3 and 4 miles) in diameter. That was when I started my golf ponderings – please, keep in mind the shape of the asteroid.

For a competent physicist, the calculation of gravitational attraction is a doddle, if you know the distance from the centre of gravity of the asteroid to the place where the measurement is taken. I knew a competent physicist, who gave me the figures, and this is where things get seriously weird.

If you stood in the middle, you would be just 6.5 km (4 miles) from the asteroid's centre of gravity, but at either end, you would be 16.5 km (10.3 miles) away, and the inverse square law comes into play. There is a major difference in the force you would feel, depending on where you stood. Walking to the end of Eros would, in fact, feel like walking uphill, because the physics of a drink-can body are not those of a tennis-ball body. You have to work to get away from the centre of gravity.

The acceleration due to gravity on the surface of Eros varies depending on where it's measured, ranging from 2.3 to 5.5 mm/s^2 (that is millimetres, not metres!) some 2000 to 4000 times smaller than on Earth. In imperial units, the acceleration would be around $^1/_{10}$ to $^1/_5$ of an inch per second per second.

A person who weighs 70 kg (154 lb) on Earth would weigh from 16 to 39 grams on Eros. Eros has an escape velocity that ranges from 3.1 to 17.2 m/s (10.2 to 56.4 ft/s), again depending on the point on the asteroid where it is measured, which would allow a golf ball hit from its surface to leave forever. That is a drive between 11 and 61 km/h (7 and 38 mph), a range well within the ability of even the rawest beginner. Throw a club away in anger, and you will never see it again!

Well, perhaps you could try gentler drives, but then comes the problem of holing out. As we have seen already, a ball on a green on Earth travelling faster than 4.6 km/h (2.9 mph) will bounce right over the standard hole. To have a ball drop into the cup on Eros, either the putt must be exceedingly slow, or the cup so wide as to make the game pointless. So the moral of the story is that golf won't be much of a sport on Eros!

the slowest flight to the moon

SMART-1 took more than a year from its launch to enter into a lunar orbit, and another two and a half months to enter a stable orbit. Blasted into space by an Ariane rocket in September 2003, the craft then used an ion-powered thruster to move out of that orbit in a slow spiral. The thruster is an experimental propulsion system that the European Space

Agency (ESA) hopes will carry probes to other planets. SMART-1 was finally captured by the Moon's gravity in December 2004

Because we only see the launch from Earth, where escape velocity is at its highest, we don't always realise that once in orbit, the rest of an escape from Earth can be gentle. The ion thruster could only push the 367 kg (809 lb) craft with a force of 0.07 newtons, a little more than the force of a sheet of A4 photocopier paper resting on your open hand. The journey took rather more than the few days taken by the Apollo missions, but at a bargain price of 100 million euros (US$120 million), ESA scientists were willing to wait.

There is a bit of a paradox about Earth orbits. A satellite travelling at 27,400 km/h (17,025 mph), 7.6 km/s (4.7 miles per second), orbits the Earth once every 90 or 100 minutes at a height of around 320 km (200 miles) from the ground. If the speed is increased though, the satellite moves out, further from the planet, into space, but it takes longer to fly round the planet. A GPS satellite at 20,200 km (12,550 miles) goes around the Earth twice each day. If you add even more speed to a satellite in an equatorial orbit, it gets further away again, and becomes geosynchronous, apparently hovering over the same place.

In other words, going faster makes you go around slower, so spacecraft like SMART-1 can just spiral around, getting further and further away until they pass a point where the Moon's gravity is stronger than the Earth's gravity.

gravity assists and wrong-way rockets

Any space mission that goes beyond Mars these days has to fly off in the wrong direction, out towards the Sun, before it heads on out to the more distant parts of the solar system, and it does so in the name of speed.

A carefully designed fly-past of another planet can give a mission a free speed boost that requires no use of rocket fuel.

The Cassini mission, launched in October 1997, gained a boost from Venus in April 1998 and again in June 1999, before swinging past the Earth in August 1999 and on to a final boost from Jupiter in December 2000. In effect, it had gone nowhere in 22 months, but after all that time, it had built up most of the speed it needed to get to Saturn, over 1 billion km (over half a billion miles) away in July 2004.

The distances might be big, but the tolerances are small: Cassini's Venus fly-bys came within 284 and 600 km (176.5 and 372.8 miles) of the planet, and the other shots were also close to the planets. Still, the basic rule of physics is that there is no such thing as a free lunch, so where does the extra energy come from? The answer is that the massive slingshot planet slows down a little bit while the tiny spacecraft gets a huge boost.

As an example, the Galileo spacecraft gained a total of 11.1 km/s (6.9 miles per second) from three gravity assists, saving 10,900 kg (24,000 lb) of fuel, 12 times what the spacecraft took off with. In a billion years, Venus will be 10 cm (4 inches) behind where it would have been in its orbit if Galileo hadn't passed by once. The heavier Earth will be about 15 cm (6 inches) behind, because Galileo had two boosts from our planet.

Michael Minovitch was a student at the Jet Propulsion Laboratory in 1961 when he saw that the principle that makes comets change their orbits when they pass close to planets could also boost spacecraft. He still invents things.

avalanches and land, rock and mud slides
Sloping piles of sand, mud, snow, rock – anything loose – has an angle of rest. It's easiest to see if you pour dry sand onto a level surface in a

thin trickle. You will see that a conical shape builds up, but as more sand collects on top, a shower of sand will run down the side of the small hill that is building up.

The angle of rest is determined by two sorts of force; gravity, which pulls every particle down, and the forces of friction, which stop the particles slipping past each other. As a rule, the frictional forces between rough particles are greater than those between smooth particles.

Around the world, there are insects called ant lions. The family or order may vary, but they have the same habit which has earned them their name. They burrow into sandy soil and flick sand away, leaving a conical hole, with sides close to unstable. Then they wait for some other insect to come blundering along over the edge. It might be a small beetle or an ant, but either way, it has the potential to be dinner, because the ant lion has found a way to catch fast prey.

As the victim scrambles up the sides, the ant lion flicks up the sand that has tumbled down with the victim. As a small boy, I learned to capture ant lions with a glass tumbler, scooping up hole and all, undercutting the ant lion by several centimetres and then sprinkling it into a tin of sand. The ant lion would soon settle in.

Over the years, I have shared this skill with others, so I have watched many times as a captive ant lion flicks sand out from under the victim, making the slope unstable, so it slides down, carrying dinner to where the ant lion's jaws wait to seize it. I'd estimate that two-thirds of the sand clears the burrow, while the other third tumbles down again, pushing at the victim.

A rock, soil or snow slope can also become unstable, though this is not usually because of any sort of undermining. The forces of friction may sometimes be reduced by the application of a lubricant such as water, either when the snow melts, the ground thaws, or rain falls. Very occasionally, the force of gravity may be changed just enough by a tremor or a vibration.

Once the material begins to move though, the previous stability is gone, because there is now another force at play, in the form of the momentum that the slope has gained. As the material moves down the slope, it pulls on the material that lies below by friction, and it weighs down upon it as well. An avalanche or a mudslide has started.

With good luck, the slide's progress will be checked by trees. In the case of a slide above the treeline though, trees won't grow (by definition), and in other places, logging may have removed this natural barrier. Thanks to friction, the speed of an avalanche will always be less than the speed the same material would acquire by falling straight down, but it can still be fast and devastating.

Snow avalanches seem to have the most potential to build up speed, and the fastest have been estimated to come downslope at 200 km/h (124 mph), which is pretty much an estimate, because nobody wants to be out front pointing a radar gun at such a thing. Still, in 1969, one Alaskan avalanche that was triggered deliberately moved at an estimated 3 km/h (2 mph). It happened at the Alyeska ski area near Anchorage, after a fall of 2 ft (61 cm) of wet snow made a slope unstable.

The snow slipped downhill, piled up on some wet snow further down, and then oozed slowly forward – still with enough force to destroy a ski lift as it came down. Sometimes even slow snow isn't safe! That said, most avalanches travel faster, depending on the slope and conditions.

A gliding motion avalanche runs between 0 and 40 km/h (25 mph). This sort of flow accelerates fast, but slides along the surface with little turbulence or mixing. The experts recommend shedding skis and poles, and making swimming motions to stay on the surface.

A dry flow avalanche runs at between 40 and 200 km/h (25 and 124 mph). Above 40 km/h (25 mph), the flow is turbulent, with a powder cloud in front of a dense core which tends to pull victims down. A strong swimming motion is the best way to stay on top.

A powder avalanche runs between 70 and 250 km/h (43 and 155 mph). This sort of flow is made up of fine particles of snow, suspended in the air. The cloud often forms when a dry flow goes over a cliff, allowing the powder to separate from the core, or when the core has stopped at the end of its run, but the frontal cloud keeps going.

A wet flow avalanche is slower, at 40 to 100 km/h (25 to 62 mph). These develop like dry snow avalanches, but they have no powder cloud. These avalanches tend to follow channels, or to be deflected around hillocks and mounds.

There is just one snag: wet avalanches are much more dense, so they are that much harder to fight against. Still, slab avalanches, where a large solid unit slides at between 90 and 270 km/h (56 and 168 mph) are even nastier because they can be 100 m (328 ft) wide and several metres (yards) thick, and that's a lot of downhill mass! Thinking about taking up swimming yet?

volcanic bombs and pyroclastic surges

Gaius Plinius Secundus, Pliny the Elder, was an enthusiastic student of nature, and also the commander of a Roman fleet. When Mt Vesuvius erupted and buried Pompeii, Pliny had some of the Roman galleys launched and taken across the bay of Naples to rescue people, but while on shore, he collapsed and died. According to his nephew, Pliny was killed by gases from the volcano, but it is more likely that he suffered a heart attack or a stroke, since none of his companions died at the time.

Still, Pliny the Elder is honoured, perhaps falsely, by modern scientists as an early martyr of science, and people have been taking risks to study volcanoes ever since. The volcanoes, for their part, have many ways of being lethal.

First, there are pyroclastic flows. These are like powder avalanches, but instead of being made of cool air and cold snow powder, the gases are fearfully hot, and the finely powdered volcanic ash is red hot. Clouds of gas and ash hurtle down, converting all life forms to instant char and cinders. The available information is sketchy, but usual speeds appear to be above 80 km/h (50 mph), though the damage to trees done by the cloud from the 1980 eruption at Mount St Helens suggest more like 480 km/h (300 mph).

Then there is the volcanic bomb, a blob of viscous lava, more than 64 mm (2½ inch) in diameter, which is flung upwards during an eruption. As it flies up and then falls, it may be pulled into an aerodynamic shape, but it also cools and hardens. The terminal velocity of a volcanic bomb is probably less than 300 km/h (186 mph), but that is still enough to do a lot of damage. Some of them end up 5 km (3 miles) from the volcano, suggesting that they took off at something more than 800 km/h (500 mph). Still, we are lucky: Jupiter's moon, Io, has a volcano that throws stuff out at 3000 km/h (1864 mph)!

a short history of long drops

Even though Leonardo da Vinci had a design for a parachute, the idea was about as useful as inventing the bicycle pump hundreds of years before the first bicycle. Even then, it would only be suited to dropping from a balloon, since it was heavy and rigid. It was too heavy even to use for jumping off a tower.

A Croatian, Faust Vrancic (or Fausto Veranzino), jumped from a tower with a da Vinci design in Venice in 1617, and survived. In 1783, with hot-air balloons all the rage, Sebastien Lenormand demonstrated the principle. Two years later, Jean Pierre Blanchard dropped a dog by parachute, and in 1793, he claimed to have jumped from a burning balloon with his new foldable silk parachute, but there were no witnesses to this. When André-Jacques Garnerin parachuted from a balloon in 1797, he had witnesses and he survived.

The first recorded parachute death was in 1837, and there were a few other cases in the nineteenth century, but jumping from aeroplanes did not take off, as it were, until 1911. Even at the start of World War I parachutes were not standard issue on either spotter balloons or aircraft. In the event of a fire, pilots and crew were left with the hideous choice of 'jump or burn'.

Garnerin's parachute had an area of 80 m^2 (861 ft^2), at the top of today's range (50–80 m^2, or 538–861 ft^2). That was lucky, as a standard parachute lands the jumper with a vertical speed of 21 km/h (13 mph). As we have seen from Figure 14 on page 159, that is like jumping from one floor up, but even a slightly smaller parachute will give the jumper a much faster vertical landing speed!

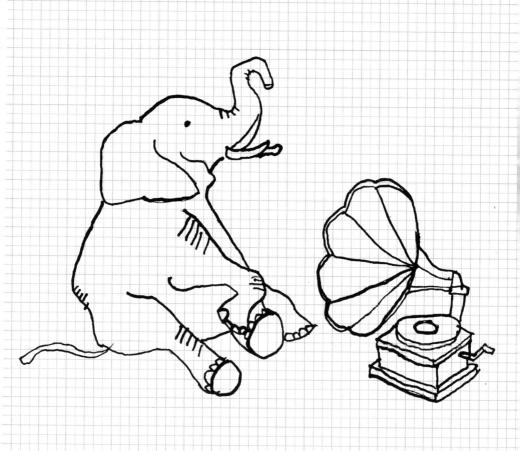

vibrate, oscillate & radiate

Back in the days of yore, when science was practised by scruffy astrologers with fleas, scabrous halitoxic alchemists with permanent bad-hair days and tattered and tarnished scrofulous mages, one heard a great deal about the 'Music of the Spheres'. The whole universe was believed to be chock-full of harmonies and vibrations that somehow held the lot together. Precision of thought was not considered essential, so long as the harmony was there.

If you were an old-time alchemist, the secret of getting patrons to put up money for wild schemes to convert cabbages to gold was usually to convince them that you, scabrous and halitoxic as you may be, had personal control of the waves, this Music of the Spheres. Of course, when the attempt failed, a good alchemist would know how to blame somebody for disturbing that harmony, making waves, as it were.

Well, the harmonies bit might be hard to sustain, but the vibrations are certainly there. They are generally neither good nor bad, but are rather the sea of energy that links the universe, or those parts they can reach.

A great deal of our life revolves around waves. The television we watch gets its signal through waves, and we see the screen because it emits light waves. Radios get their signals from waves and emit sound waves that we can hear.

Even a CD player uses light waves from a laser to detect a signal burned into the CD, and then from that, generates sound waves. We can cook our food in a microwave oven or in an ordinary oven that emits heat waves, and three of our senses are attuned to waves because we can detect light, sound and heat.

sound in air and other media

The speed of sound in air depends mainly on the speed of the gas molecules that carry the pressure wave to the listener. At standard room temperature, 25°C (77°F), oxygen atoms are travelling at 482.1 m/s (1581.6 ft/s) on average, 1735 km/h (1078 mph), much faster than the speed of sound.

Close to sea level, there isn't a lot of free space between molecules. In fact, the average air molecule only travels around 0.1 μm, a tenth of a micrometre, one ten-millionth of a metre (metre = 1.094 yd) before it hits another molecule. The energetic molecule that is carrying a small part of the pressure wave then hits a second molecule, and transfers its energy. Gas molecules are not billiard balls or ball-bearings though, so the collisions themselves slow things down. On top of that, the collision may send the next molecule in a slightly different direction. The result is that sound travels slower than the gas molecules.

In the end, all of these uncertainties average out, so the speed of sound in dry air at 0°C (32°F) is 331.3 m/s (1087 ft/s). If you warm the air up to 20°C (68°F), the speed of sound in air is 343 m/s (1125 ft/s). At 100°C (212°F), it is 386 m/s (1266 ft/s).

There is a rough formula for calculating the speed of sound in metres per second at different temperatures: $V = 331 + 0.6T$, where T is in degrees Celsius. The standard value for 20°C (68°F) fits this fairly closely, while it provides a value that is slightly high for 100°C (212°F).

When it comes to liquids and solids, the speed of sound is much greater. In sea water, sound travels at 1500 m/s (4920 ft/s), 5400 km/h (3355 mph). In soil, sound travels at more than twice that speed, and it travels in steel at 5100 m/s (16,732 ft/s), 18,400 km/h (11,433 mph). As materials get denser, sound travels ever faster, and that becomes important when we study earthquakes.

marin mersenne and the big string

Most small children discover what happens when you stretch several rubber bands over an empty margarine tub or some sort of similar box. To a three-year-old, this makes an eminently satisfactory guitar, though to more developed musical tastes, it leaves something to be desired, if only because tuning is very difficult.

All the same, the basics are fairly obvious: a tighter string makes a higher note, and so does a shorter string. It takes a bit more sophistication to work out that a thinner string with the same tension and length also makes a higher note than a thicker but otherwise similar string. It takes a lot more sophistication to work out that the string is vibrating backwards and forwards, and that the higher the note, the faster the vibrations.

Pythagoras, the Greek mathematician with the well-known theorem, was the first to investigate this, but a proper study of musical notes had to await the efforts of Marin Mersenne, an enthusiastic gossip about science in an age when communication was slow and disorganised, and also a keen experimenter.

Mersenne got around the problem of measuring high notes by measuring very low notes. His plucked strings were a hemp rope more than 30 m (98 ft) long, and a brass wire 43 m (141 ft) long. On both, the vibrations were so slow that Mersenne could see each individual wave. By varying the length and the tension on these giant strings, he was able to derive a formula that tied the frequency of a string to the length, the tension and the mass of 1 m (3.3 ft) of the wire or string.

After that, it was a simple matter to set up a short wire or string, listen to the note, and calculate its frequency, which could then be compared with more portable standards, like tuning forks.

standard pitch

Standards are not always all that they might be. Around 1853, 19-year-old Johannes Brahms toured northern Germany with a 23-year-old Hungarian violinist, Eduard Remenyi, giving recitals. Legend has it that on one occasion, the piano Brahms was to play had been tuned half a tone too low. Rather than spoil the effect by having Remenyi tune down his violin, Brahms undertook to transpose the *Kreutzer Sonata* on the fly, and did so to great effect.

These days, instruments are generally (but not always) tuned to having the A above middle C equal to 440 Hz, but this was not always so. In one rare case, an A of 380 is known, as is an A of 480. Once tuning forks were invented in the early eighteenth century though, there was an opportunity to get something closer to a standard.

In the middle of the nineteenth century, the issue began to be taken seriously, especially by singers who could find one of their favourite works impossible, because the accompanying instruments were off on a scale of their own. The speed at which air vibrated in string instruments could be adjusted with ease, wind instruments were more of a challenge, but singers' throats had a fixed range and were an entirely impossible situation.

Even today, with electronic systems having been perfected, much of the United States and Britain is happy with A440 as a standard, while other US orchestras use A442, and parts of Germany, Austria and China prefer A445.

Then again, just as with the oldest wind instrument, the dijeridu, temperature is also going to play a part!

the well-temperatured dijeridu

When the air inside a concert hall gets warm, the string instruments sound slightly flat because the strings expand as they warm, making them lose a little of their tightness. With the wind instruments, on the other hand, it goes the other way: in warmer air, sound travels faster – a note played by a wind instrument depends on the time it takes for a sound wave to get from one end of the tube to the other.

This means that when the air gets warmer, the notes for a particular tube in a fixed setting become sharper than they were. This even applies to a dijeridu, the traditional musical instrument of some Australian Aboriginal peoples.

A proper dijeridu is complex, because it is a slim log that has been hollowed out unevenly by termites. This makes for difficult calculations, but we will just take it as a smooth bore in this treatment. I am also leaving out a few complications like the end effect because it doesn't change this story, but now you know what is missing, you can look it up.

If we use (v) for the speed of sound, then the natural frequency (f) for a tube of length (l) is given by $f = v/4l$. If you want to have a dijeridu that is 1200 mm (about 4 ft) long, what frequency will it have?

At 20°C (68°F), with the speed of sound 343 m/s (1125 ft/s), the frequency will be 343/4.8 Hz, which comes down to 71.46 Hz. On a cold clear night in northern Australia, the speed of sound is 331.3 m/s (1087 ft/s), which equates to a frequency of 69.37 Hz; on a hot day at noon, the speed of sound at 40°C (104°F) is 355 m/s (1164 ft/s), which will turn up a frequency of 73.96 Hz. The difference of 4.6 Hz is equivalent to going from C-sharp to D, a full half-tone – and all depending on the weather!

fast music

At the age of 17, Mozart composed his 27th symphony, all in a single day, but at least he never had to put up with a sheet music publisher giving it a bizarre nickname! Later composers hated the nicknames given to their works, but it is likely that few would have as good a case as Frédéric Chopin with his Waltz in D flat Major, Opus 64, No. 1, known around the English-speaking world as 'The Minute Waltz'.

From a sampling of recordings, it appears that the usual playing time is between 1½ and 2 minutes, and those who try to play it in less are usually dismissed as 'musical gymnasts'. Artur Rubinstein recorded it at 1:48 and Vladimir Ashkenazy at 1:49, and other times up to 2 minutes are not unknown.

Legend has it that Chopin was inspired by the sight of a small dog chasing its tail, and that he called it by the French title 'Valse Minute', meaning the little or tiny waltz – time was not involved. All the same, many performers have managed to get through the work in less than a minute: probably the fastest ever was Liberace, who could finish the work in just 37 seconds, and often did.

Sergei Rachmaninoff made an arrangement of 'Flight of the Bumblebee' and cut an Ampico piano roll that lasts a fraction more than a minute. While Ampico rolls are open to doctoring, accordionist Alexander Dmitriev has been filmed playing this work on a bayan (Russian chromatic accordion) in just over 65 seconds by my timing. Accordionist Liam O'Connor, acclaimed by the *Guinness Book of Records* for achieving 11.64 notes per second, could probably do better.

Gabriel Fauré was once asked how fast one of his songs should be sung. He suggested that if the singer was notably deficient of talent, it should be sung very fast indeed!

slow music

Few pieces of musical composition can have an accurate time attached to them. John Cage created '4' 33"', a piece of absolute silence which lasts exactly that long. Most classical music station engineers loathe it, because networks typically have a system that plays standby music after the transmitter has been silent for 90 seconds, and the very nature of Cage's work ensures that this fallback is triggered at least once (and often twice) in a playing.

By an odd chance, Cage also brought fame to perhaps the longest piece of music ever conceived. Erik Satie produced 'Vexations', his marathon piece on a single sheet in 1893, and Cage arranged the first complete performance in 1949. There is a catch: while the music can be written on a single sheet, the theme is to be played 840 times, 'very slowly'.

Another piece of French music, from Poulenc's 'Dialogues of the Carmelites', only seems excruciatingly slow. A chorus of nuns sing 'Salve Regina' as they file off stage to their deaths, with the interpolation, every so often, of the swoosh of a guillotine blade falling. Each time the sound is heard, one of the performers stops singing. To listeners, the chorus seems to go on for hours.

The guillotine, on the other hand, is very fast. It takes less than one second for the 40 kg (88 lb) blade to drop 4.3 m (14 ft), by which time it is travelling at just over 9 m/s (29.5 ft/s), 33 km/h (20.5 mph), severing the neck in two one-hundredths of a second. Death may take a few seconds, then all is silence.

Some pianists may be tempted, after trying to play 'Vexations' as a solo work, to wish they had lost at least their fingers to Madame la Guillotine. At least they would still be able to play '4' 33"'.

ernst mach and the supersonic

When you hear sound from a radio, pressure vibrations are coming from the loud speaker as the speaker diaphragm vibrates several thousand times a second. Each time the diaphragm moves towards you, a pressure wave comes your way. And each time the diaphragm moves away from you, a non-pressure wave comes your way. A rarefaction, it should be called, but they're more commonly known as compressions and decompressions, pressure and non-pressure.

When we speak, we generate vibrations in the same way, while a clap happens when we trap air between our hands, causing a sudden pressure wave. The same thing happens when we burst a balloon or a paper bag, or let off a gunshot or a firecracker, as they all involve a single pressure wave. Thunder is complicated because lightning makes lots of small compressions at different distances, so we hear a rumbling series of small random pressure waves as they reach our ears.

The compressions and decompressions come racing at you at about 330 m/s (1082 ft/s). So why don't they knock you flat on your back? The simple answer is that there just isn't a lot of energy in the sound waves. There are only a few molecules actually involved, and the pressure difference between the height of the compression and the depth of the decompression is quite minor.

Curiously, the speed of sound is independent of the speed of the source. If the source is moving, we experience the Doppler effect, explained in 'Doppler radar' on page 16. Imagine that you are you are on a boat travelling at 60 km/h (37 mph) and that you throw a tennis ball at 30 km/h (18.6 mph). If you throw the ball forward, it will have a combined speed of 90 km/h (56 mph), as seen by a stationary observer. But no matter how fast you go, the sound you make travels out from you at just the speed of sound, no more and no less, and in all directions.

So a plane flying close to the speed of sound is travelling almost as fast as the pressure waves of the sound that it is making. These are small, insignificant pressure waves that wouldn't hurt a fly. Their individual effects are no worse than being hit with a piece of straw. But there was once a camel who had to bear the brunt of that last straw, wasn't there? And that's the problem when you start to pile pressure waves up on top of each other: you get sonic booms. A plane which approaches the sound barrier gradually is likely to break up from the buffeting it receives from the pressure waves, so like somebody entering a cold swimming pool, it has to crash through.

You might think the problem of air flow at the speed of sound was one that was only noticed when jet aircraft started to fly close to the sound barrier. In fact, the effect was known much earlier, but until jet planes were common, only specialists knew or cared very much about the sound barrier: it was a rather theoretical sort of thing.

Back in 1884, Ernst Mach took photographs of the shadows sound pressure waves make, and within a few years, he announced that there were changes in the air flow over objects moving at the speed of sound.

'Mach numbers' were first referred to in the mid-1920s, probably about the time when the propellers on aircraft went round so fast that their tips approached the speed of sound. This effect, by the way, imposed a limit on the top speed of propeller-driven aircraft.

ernst mach's amazing shadowgraph

Ernst Mach published his first paper on supersonic movements, complete with photographs, in 1887, in Vienna. The supersonic object he studied was a bullet, but it was the photograph that amazed physicists.

The picture shows a bullet travelling from right to left at a supersonic speed. A cone of pressure is spreading out to either side, and a small trail of turbulence follows along behind. The bullet and its trail are viewed through a window in the apparatus. The bullet is lit by a bright point-source spark of light. The challenge is to work out how Mach managed to photograph a transparent pressure wave in air, which is transparent as well.

Anybody who has snorkelled into a freshwater spring coming up through saltwater will have encountered a peculiar blurring as the more dense salt water and less dense freshwater mix. As light passes from one medium to another, its speed changes slightly and it is bent, causing the blurring.

The more dense air in the pressure wave is also hotter than the surrounding air, and light from the spark source is bent, and that is what causes a shadow to be thrown onto a white screen on the other side of the viewing window. So what we see in the photograph is a trick, a bit of a fake: we see an image of the bullet, combined with the shadow of the pressure wave on the screen behind. It was brilliantly, deceptively, simple.

the cameras that lie

Just as Ernst Mach was able to capture the movement of a bullet with a fast camera, so too do scientists and engineers use time-lapse photography to 'fast-forward' slow actions, or slow motion photography, to allow for very fast action to be examined in detail. At times, these effects are used for less honest reasons.

Car chases in movies were often filmed at low speed, using cars loaded with sandbags or other weights. Later, the film was shown at 24 fps, and the cars, which cornered realistically, thanks to the sandbags, appeared to be going much faster. In the same way, martial arts movies are still often shot at a speed lower than the usual 24 fps. When they are then shown at 24 fps, the action appears amazing.

When the martial arts star Bruce Lee was first film-tested to see how he performed, a problem arose: he was just too fast. Many of his action scenes had to be shot at 32 fps – this captured the action so that it could be seen when the film was projected at 24 fps.

Stop-frame photography that speeds up slow action, like slow motion replays of sporting events, have made us more familiar with these effects, but there are still hidden wonders.

The fastest camera available to science at the moment takes 200 million fps, enough to capture really fast things breaking apart. Commissioned in 2000, it replaced a camera that could only take 800,000 fps, meaning that some of the studies that now take a fraction of a day took several days to complete. Potential applications proposed at the time included the damage caused by explosions and the effects of bullets on Kevlar vests.

a fast history of the speed of light

We now accept a value for the speed of light in a vacuum of 299,792,458 m/s (327,857,018 yd/s), or for almost all purposes, 300,000 km/s (186,000 miles per second). This value is fixed forever, because the metre is now defined in terms of the speed of light. In the past, a variety of values were quoted.

The first recorded attempt to measure the speed of light was made by our old friend Galileo. Before him, a clever technician called Hero of Alexandria had a try. Hero thought that light came *from* the eyes, bounced off something and came back, a bit like Superman's 'X-ray vision'. He knew the Sun was a long way off, but as you could see the Sun as soon as you opened your eyes, the speed of light, he thought, was infinite.

Galileo knew that light came from lanterns to begin with, so he had people on two distant hills flash lamps back and forth at each other, with an observer uncovering his lantern in reply to a flash from the other hill. In a quiet place, you might use shouts and replies to estimate the speed of sound in this way. Sadly, the time taken for light to travel between the hills they used was only about one forty-thousandth of a second, so once again, the answer was that the speed of light *seemed* to be infinite.

Later in the 1600s, a Dutch scientist, Christiaan Huygens, pointed out that during a lunar eclipse, the Sun, the Earth and the Moon appeared to line up very closely. That implied that the Earth's shadow had to reach the Moon in 10 seconds or less. He suggested that the speed of light might be 100,000 times the speed of sound, which would be 33,000 km/s (20,500 miles per second), about 11% of the value we accept today. It was a start: now light was fast, but at least of finite speed.

And then in 1675, Ole Rømer, a Danish astronomer, noticed an oddity. It concerned the eclipses of Jupiter's four large moons (discovered, as you may have guessed, by Galileo). At times, they sped up for about 6 months, then they slowed down for about six months, before speeding up again.

Rømer knew about the way the solar system was laid out, and he guessed that the difference had something to do with whether the Earth was on the side of its orbit nearest to Jupiter, or on the side furthest from Jupiter. The only problem was that he had no idea how big the Earth's orbit was. He had to guess, and ended up with a speed of light of 227,000 km/s (141,050 miles per second), about 76% of the modern value.

In 1728, an English vicar and astronomer, James Bradley, used the same method to reach a value of 283,000 km/s (175,850 miles per second and 94% of the true value). That stood until 1849, when Armand Fizeau improved on Galileo's method. He shone light through the gaps of a toothed wheel, at a mirror, 8 km (5 miles) away. If the wheel spins fast enough on this apparatus, the light gets back, just as the next gap moves into position. If you know how many gaps there are in the wheel, and how fast the wheel is rotating, you will have a fairly accurate estimate of the time taken for light to travel 16 km (10 miles). Fizeau's result was about 5% higher than the value we accept today, but the value was adjusted downwards in the following year, when Jean Foucault refined the method even further.

In 1870, James Clerk Maxwell, a Scot, deduced that light had to have both an electrical component and a magnetic component, so light had to be a form of electromagnetic radiation. In 1888 in Germany, Heinrich Hertz set out to generate an electromagnetic wave by passing an alternating current over a spark gap. Then he set out to measure the wavelength of that wave.

Today, the old-fashioned among us might call that a Hertzian wave, but most physicists just call it a radio wave. Whatever the waves were called, they could be reflected like light, they interfered with each other like light, and they could be concentrated by a concave steel mirror. Hertz also showed that the waves travelled in a straight line, were polarised and could be refracted, just like light. (Refraction is the form of bending of light that we see in a lens or a prism.)

In the interference study, Hertz found bright spots at 33, 65 and 98 cm (13, 25.5 and 38.5 inches) from the source, suggesting a 33 cm (13 inch) half-wavelength, and '1.1 thousand-millionth of a second as their period of oscillation, assuming that they travel with the velocity of light' – and from his other findings, he had little doubt they did.

When people talk about Einstein, the one thing everybody knows is '$e = mc^2$'. People are always amazed that c^2, the square of the speed of light, enters into consideration. Einstein, they conclude, must have been a genius, and he undoubtedly was, but not for introducing c^2 into physics.

When you read the original physics research of the period at the end of the nineteenth century and the start of the twentieth, that mysterious little c^2 appears a number of times, yet nobody really knows why c has come to stand for the speed of light in a vacuum. It just does. Some people say it is short for 'celerity', a fancy Latin-derived word that means speed, but it doesn't really matter.

Then again, what about the speed of gravity? According to Einstein's general theory of relativity, gravity travels at the same speed as light. But, nobody has ever detected gravity waves, and it seems that the only way you could assess the speed of gravity would be to create a large mass where there wasn't one a moment ago. Until now, at least, that sort of trick has been beyond physics.

slow light and making spectacles out of rainbows

One of the few things all people take from their science education is that the speed of light is constant, that it is always c, about 300,000 km/s (186,000 miles per second). It can, of course, be measured in other units. You could also say for example that the speed of light is about 1.8 trillion furlongs per fortnight (1,802,617,498,752, if you want precision), but even though you can say that, you'd be wrong. You are more wrong now, but you have always been wrong.

What is constant is *the speed of light in a vacuum under normal conditions*. The speed of light in other media is quite variable. Just to really spoil your day, when light passes through a window, it slows down, and then it speeds up again when it leaves the glass, a bit like vehicle traffic passing through a toll plaza (try not to ask how the light accelerates: it hurts the brain).

We use something called the refractive index (we will see why in a minute) to indicate the way light is slowed down. A refractive index of 1 means no slowing, a refractive index of 2 means the speed is halved, and index of 1.5 means $2/3$ of the vacuum speed, and so on.

Here are some examples: air (1.0003), ice (1.30), water (1.33), glycerine (1.47), linseed oil (1.48), microscope immersion oil (1.515), glass (1.52), flint glass (1.66), zircon (1.92), diamond (2.42), lead sulfide (3.91).

One effect of this slowing down is that light that approaches the new medium at an angle is bent. There is a scientific law, Snell's Law, that links the amount of the bending to the differences in speed in the two media. We call the bending refraction, and the refractive index (n) of a transparent material is given by n = c/v, where (v) is the speed of light in the material.

Refraction explains the way a pencil or a drinking straw appears to bend when it is poked into a glass of water. Refraction also explains why lenses work as they do: if you use spectacles to read, you are relying on the speed of light being changed, close to your eyes.

Refraction, the varied speed of light, is also behind the beauty of the rainbow. Remember Isaac Newton? He discovered this one as well, or at least he described it first. He could quite possibly have been inspired by somebody else, because the way he explains his experiment, he clearly already knew what needed to be done. It boils down to this: light of different colours, different wavelengths, is slowed down by different amounts, and that means it gets bent differently.

A rainbow is always in the sky furthest from the sun – it is formed when light is reflected and refracted several times inside a raindrop, and the colours get reflected in different directions. You see red in the part of the sky that is reflecting red light towards you, yellow in the part reflecting yellow, and so on.

Pavel Alekseyevich Cherenkov's name became well known during the earliest days of space flight, when it was reported with some excitement that astronauts and cosmonauts had seen flashes of light. These flashes, we were told, came from particles that were slowing down to the speed of light, but few of us were fooled by that. We all knew nothing travels faster than the speed of light.

Forewarned as you have been, you might also see where this is leading. When a particle travelling at a velocity close to c enters a more dense medium, it is travelling faster than light does in that medium. In glass, the limit is about two-thirds of the vacuum speed, and in water, it is about three-quarters of the speed of light in a vacuum. The eye is largely filled with water.

In slowing, the particle radiates energy as bluish radiation and slows down. The speed of the fast charged particles can be estimated by measuring the angle at which the Cherenkov radiation is emitted as the particles pass through the transparent medium.

Previous researchers dismissed it as just fluorescence, which happens when something absorbs radiation at one wavelength, and emits it at another wavelength. In the 1930s, Cherenkov began a study of the ways different liquids absorbed the radiation from a radium source and during this process, he saw a bluish glow. Unlike others, he did not accept that this was a simple fluorescence radiation which should be sent out equally in all directions. He noted that the blue light that he was studying was directional, which meant this was something entirely different.

Cherenkov tried adding some standard chemicals which quench fluorescence, but nothing changed. When it turned out that the light was polarised along the direction of the incoming radiation, but in quite a different way from polarised fluorescence, it became even more interesting.

The final clue came when Cherenkov found that the radiation is only emitted at a certain angle to the radiation that gave rise to it, like the wash of a boat. In the end, Cherenkov did not explain the phenomenon, but two other researchers, Ilya Frank and Igor Tamm, did. They shared the Nobel Prize with him in 1958.

Then, a few years later, people up in space reported seeing flashes of light and physicists realised that this was Cherenkov radiation, formed inside human eyeballs, as high-speed cosmic particles whizzed through them, faster than the speed of light inside the eyeball. If you see film of a nuclear reactor where radioactive material is kept under water, look for the eerie blue glow. This is also Cherenkov radiation, caused when particles emitted by the fuel and travelling faster than the speed of light in water are slowed down.

earthquake waves

Once upon a time, earthquake waves were just shakes, mostly of unknown origin. People who lived near volcanoes could see that a few earthquakes were associated with volcanic activity, but the correlation was hard to spot.

For example, there are earthquakes but no active volcanoes in seismically active Turkey, though there are plenty of volcanoes dotted around other parts of the Mediterranean – and Australia, Pakistan and Portugal all have occasional earthquakes but no volcanoes at all. It was easier to propose that a giant or a dragon under the Earth was stirring uneasily in its sleep. It was far more realistic at the time, and easier to believe than the truth, which is that our safe bits of land, dubbed *terra firma* by the Romans, were far from firm. In reality, most of the bits we live on are moving at a sedate pace, about as fast as a fingernail grows. The Earth's crust floats on a hot mantle layer.

Where the plates rub past each other, they grip and stick, they flex and tense up, and in the end, they slip again. Think of two small children, tugging on a rubber band: when it breaks, somebody is going to be hurt as the energy stored in the rubber is forced to relocate.

That energy is what we call an earthquake. You can get a lot worse than a simple earthquake though. All good things come to an end, and at some point, the plate that moves forward has nowhere else to go. When that happens, it must plunge down under some other plate, carrying all sorts of silt, sediment and water with it. An instability has been introduced into the Earth, and sooner or later, there will be a price to be paid for adding all that lubricating water.

We call these regions subduction zones, and it is at these areas where the world's largest and most dangerous earthquakes take place. The main subduction zones are in Japan, Alaska, Mexico, Central America, Peru and Chile, all recognisable as earthquake centres. Of course, to some people, earthquakes are remarkably enlightening, thanks to the waves they generate, and the speeds at which they travel.

To get an idea of what these seismic waves look like, try pushing one end of a Slinky spring that is hanging from threads. What you will see is a compression wave moving along it. If you stretch a long rope and flip the end of it, you can see a wave of up-and-down motion travel its length. You can make a whole career out of studying the various waves that come from earthquakes, but for this discussion, I will stay with those two types. For experts, you'll note that there is no mention of Love waves or Rayleigh waves here (I didn't forget, I just chose not to).

Seismic waves travel some kilometres (miles) each second. The actual speed depends on the type of rock they are travelling through, because like sound waves, seismic waves are affected by density. It also depends on the type of wave. The first waves to arrive are called primary waves (P), while the later ones are, as you might expect, called secondary waves (S). Now imagine that two cars have started from the same unknown starting place. One travels at 50 km/h (31 mph), the other at 100 km/h (62 mph). If the secondary car passes your house one hour after the primary car, how far away from you did they start? I leave it to you to work out why the answer is 100 km (62 miles) away.

The same sort of calculation can be applied to waves. In the range of 50 km (31 miles) to 500 km (310 miles) from the surface, P waves travel at about 8 km/s (about 5 miles per second), while S waves travel at 3.45 km/s (about 2 miles per second). The mathematics is more complicated, but we can draw a circle marking all the places at the right distance to the origin. If we add several more circles from other stations, we see an intersection and know where the quake happened.

Some of the compressional waves travel through the centre of the Earth, and as they pass through rocks of different densities, they are refracted. With clever mathematics, we can use this information to learn about the inner workings of the planet. Maybe earthquakes aren't so bad!

tsunamis and tidal bores

Tsunamis were always called tidal waves until 2004, when a disastrous quake on 26 December killed more than 300,000 people in those parts of Asia facing the Indian Ocean. It was the first major tsunami of the Media Age, and by the time its carnage had become old news, the world had acquired a new word.

In hindsight, the risk of a tsunami was obvious. A huge earthquake, south of Australia, indicated a major shift in the crust of the Earth, but it scarcely caused a ripple because the plates moved sideways. It shifted tension from way south of Australia to the north, where the north-pushing

Australian plate would be more likely to involve a vertical movement and cause chaos.

Hindsight tells us it was likely to happen soon, but giant tsunamis are rare enough for people not to think of them. They are caused when something large, like an asteroid or a cliff falls into the sea, when a major slide of material happens under the sea, or when the ocean floor heaves after the crust breaks.

In the quake in 2004, the 1200 km (745.6 miles) rupture opened lengthwise to the north-northwest at 2.5 km/s (1.5 miles per second) in the first 10 minutes of the earthquake. There was also a Doppler shift in the waves, with seismometers to the north, in Russia, recording a higher frequency than those to the south in Australia. More than 30 km^3 (18.5 mile^3) of water was displaced by the shifting sea floor, and this generated the waves that did the damage.

The Atlantic coasts of Europe, Africa and the Americas are at risk from the Canary Islands, where a future eruption of the Cumbre Vieja volcano could trigger an undersea collapse, with a block of rock '... twice the size of the Isle of Man ...' set to fall at up to 350 km/h (217.5 mph). The energy released would be equal to the electricity used in the United States in 6 months, and it would mostly go into the water.

The name tsunami comes from Japanese, and means 'harbour wave', because boats out to sea will generally feel and see no wave at all, but close to the shore and in harbours, all the energy of an 800 km/h (500 mph) wave is used to form a much higher but rather slower wave. There are also problems when a harbour manages to focus the effects of a tsunami.

But why use a Japanese word? Japan is largely volcanic, sitting on a subduction zone, and has long been a literate society. They were, and are, aware of tsunamis, all too painfully aware. It's a good word, just a bit hard to pronounce.

The first result would be a dome of water, 900 m (2953 ft) high and tens of kilometres wide, generating a series of crests and troughs, surging out at 800 km/h (500 mph). The waves spread, losing energy with distance, so the 100 m (328 ft) waves on the western Sahara shore, would be 50 m (164 ft) waves in Florida and the Caribbean, eight to nine hours later. The waves hitting Brazil could reach 40 m (131 ft), but in some places, the waves could be funnelled, as they were at Hilo in Hawaii in 1960, when 61 people were killed and 282 injured. The only real certainty is the speed of transmission.

Sooner or later, an asteroid will touch down in an ocean. Asteroid 1950 DA is 1 km (0.6 miles) across, and there is a slight chance of it hitting in 2880, but there are probably others, still to be spotted. An asteroid that size would blast a cavity in the Earth about 18 km (11.2 miles) across, and it would reach all the way down to the sea floor. As the ocean rushes in, waves will radiate out at 800 km/h (500 mph), and rather than the one big wave that movie-makers like, there will be a number of them, starting small and getting larger, one every three or four minutes, or sometimes more.

This variation is important if you are ever caught in a tsunami zone: in Sri Lanka in 2004, the biggest crest was the third or fourth, which gave a British geologist a chance to warn hotel staff and tourists to clear the beach. In the ocean, those waves were only a metre high, but their tremendous speed converted to extra height in shallow waters. The wave system also leads to a 'drawing-down', where the sea appears to go out: if you see this, run!

Sometimes, though, there will be no warning. In July 1998, a landslide caused by a magnitude 7.0 earthquake triggered a tsunami on the north coast of New Guinea, sending water surging in, just minutes later, at 10 to 20 m/s (33 ft/s to 65.5 ft/s), about 30 to 60 km/h (19 to 37 mph). A wind at the top end of that scale can buffet you, but the force of a water wave is about a thousand times as strong as a wind of the same speed.

quantum lite and quantum light

A word of warning up front: according to quantum physicist Richard Feynman, 'nobody understands quantum physics'— and this is about quantum physics. It's like the black hole stuff: just take it in, nod agreeably and don't sweat.

According to quantum physicists (that is, people who think they understand quantum physics), you can stop light dead in its tracks, capture it for a while, and then let it go to streak off again at the speed of light. What's more, they have been able to publish proof of this in *Nature*, one of the world's most prestigious peer-reviewed journals of science, which means it has been carefully assessed and accepted.

In 2003, a group of researchers reported what they called frozen light, meaning light that gave every evidence of having been slowed down as it passed through an ultra-cold gas, to 17 m/s (56 ft/s). The ultra-cold gas, they said, was handy in helping measure the speed of the light, and that they had reduced light to speeds of just under 100 m/s (328 ft/s) in a gas made of hot rubidium atoms. Nod in agreement, OK?

Where will this lead? Nobody can really say, but back in 1888, when Heinrich Hertz was measuring the waves emitted by a spark gap, who could have predicted that there would one day be radio stations that needed, because of a copyright dispute, to play a lot of music by people who were not the usual singers of popular music, people who sang what we now call rock and roll?

In short, the discovery that light can be slowed right down is just one of those things where you have to nod agreeably and move on – if a copy of this book exists in a century's time, somebody can have a giggle, wondering why I failed to see where it was going. Nod agreeably!

78 = 33 + 45

When music first became available on disc, the grooves on the discs had to be coarse, because the needle that traced a path along the spinning groove was also coarse. The standard was set around 1900 for a 25 cm (10 inch) disc of shellac, to be spun on a turntable at between 70 and 90 rpm.

The average, 78 rpm, was adopted after 1912, although a few records were cut at 80 rpm even in the mid-1920s. Clockwork phonographs (or gramophones) were fitted with a mechanical governor, allowing them to deliver a wide range of speeds, all carefully controlled.

Sophistication crept in slowly during the 1930s and 1940s, but the established base of players maintained the 78 rpm standard, and with electric turntables, it was difficult to vary. The 12 inch (30 cm) 33 rpm LP was invented in 1948, and most discs were made of vinyl as it was more flexible than shellac. The grooves were far smaller, giving the name 'microgroove' to these new discs.

The 45 rpm single-track vinyl disc was invented in 1949, and the age of the 78 was nigh. The standards were largely based on somebody's priorities or guesstimates about what would work. The 33 was actually $33^1/_3$ (one third of a hundred), which was a good choice, because strobes based on either 50 Hz or 60 Hz mains power could be developed to test that speed.

Mathematically, a disc record of constant rotational speed works best when the inner recorded diameter is half the outer recorded diameter. This determined that a 7 inch (18 cm) disc would have a blank inner that was 3½ inches (9 cm) across, and from there, given the groove width, song lengths and a few assumptions about quality, the 45 rpm standard was born, and Mr Presley had a medium to exploit.

There was also another rather less common LP standard that was used mainly for recording speech. In this standard, the disc turned at $16^2/_3$ rpm, but it was little used. There was, in general, over time, a continual fall in the rate at which discs turned on the turntable.

The opposite, however, has happened with hard discs, compact discs and DVDs, where the trend has generally been to ever higher speeds and denser storage, allowing much faster access times.

The first popular floppy disc, the 5¼ inch (13.33 cm) disc, was spun up to either 300 or 360 rpm, while the later 3½ inch (8.9 cm) floppy discs turned at 300 rpm. These discs are magnetic, so they were read by a magnetic head sitting close to the surface of the disc, picking up digital signals – data, images, music or something else.

A CD spins at a variable speed that delivers linear tracking at the rate of 1.2 to 1.4 m/s (3.9 to 4.6 ft/s), translating to somewhere between 200 and 500 rpm, depending on which part of the CD is being accessed. DVDs similarly rely on a constant linear velocity, and so they rotate at between 570 and 1600 rpm. The CD and the DVD, however, rely on a small laser that detects tiny pits in the surface of the disc; these pits spell out the stored digital signal.

Hard disc drives on the other hand whirl at between 3600 and 15,000 rpm, and once again we are back to magnetic reading.

By comparison, neutron stars spin at somewhere between once every 16 milliseconds and once every 8.5 seconds, giving a range somewhere between 3750 and 7 rpm. This high speed arises because neutron stars shrink as they form, and like a skater who pulls her arms inwards, this makes them spin faster.

the voyager spacecraft whiz to the edge

Somewhere around 16 December 2004, the Voyager 1 spacecraft passed into a border region at the edge of the solar system. That was when the craft crossed the termination shock that marks the start of a transition region at the edge of the Sun's heliosphere called the 'heliosheath'.

To us, the Sun is a fixed point, but in reality it is pushing along around the Milky Way galaxy at 900,000 km/h (560,000 mph), surrounded by the heliosphere, a bubble maintained by the flow of charged particles that we call the solar wind, which pushes interstellar dust, charged particles and gas out of the way.

The termination shock is a shock wave in the solar wind, marking a massive slowing of the solar wind at the point where the outside pressures shove back, slowing the solar wind down. Some time around 2014, Voyager 1 will reach the heliopause, which is the end of our solar system. At the heliopause, the solar wind is completely stopped by the pressure from the interstellar clouds.

Both Voyagers 1 and 2 are capable of returning scientific data from a full range of instruments and they have enough electrical power and attitude control propellant to keep operating until 2020. Voyager 1 is now drifting at 17.374 km/s (10.795 miles per second), 62,546 km/h (38,864 mph), while Voyager 2 is travelling at a more sedate 15.957 km/s (9.915 miles per second), 57,445 km/h (35,695 mph), relative to the Sun.

Voyager 2 should cross the heliosheath somewhere around 2008. The two craft have all the power they are ever going to have. Each carries a golden gramophone record with information about the beings who made the craft. Who could have predicted that before the craft even stopped working, that sort of record would be ancient technology?

sunjammers give way to plasma beams

Science fiction has long favoured the idea of a spacecraft that somehow made it into orbit and then sailed away, powered by the pressure of the solar wind on gossamer sails. The SF writers changed the old name for sailing ships (windjammers) and dubbed them sunjammers.

One of the extensions of this idea was to make a plasma bubble serve as a sail. Robert Winglee has since pushed this idea to a new level, suggesting that plasma beam generators in space might be able to offer a return trip to Mars in 90 days: right now, the best technology could only produce a round trip in about 30 months, 900 days.

A faster trip is attractive, because it reduces the chances for things to go wrong when solar storms come along, or just when ordinary medical emergencies arise. The idea is to place one or more space stations along the way, generating plasma beams to push the spacecraft along. Winglee believes that a control nozzle 32 m (105 ft) wide would generate a plasma beam that could push a spacecraft at 11.7 km/s (7.270 miles per second), 41,842 km/h (26,000 mph), or 1,005,840 km (625,000 miles) a day.

Mars is some 77 million km (48 million miles) away, on average. At that speed, a one-way trip would take 76 days, but Winglee hopes to shave this to deliver a 90-day return trip. The big advantage is that the propulsion unit will be left behind, so much less mass has to be accelerated. The catch is that a second propulsion unit must be at the other end, to slow the spacecraft down again.

He suggests placing units for various missions and extracting multiple uses from them. Nearby ones could use solar panels to generate electricity, more distant ones might use nuclear power. And unlike the sunjammers, these units could be accelerated, even past the heliopause.

the howling solar wind

The solar wind is a stream of plasma from the Sun that rushes through space until it hits the Earth. One part, the 'low speed' solar wind, travels at a mere 1.5 million km/h, or 1 million mph (about 400 km/s, or 248.5 miles per second), while the high speed wind travels at twice that speed.

The SOHO (Solar and Heliospheric Observatory) was launched by NASA and the European Space Agency in 1995. It was only planned to operate until 1998, but was still working in 2008. SOHO orbits around Lagrange Point L1, 1.5 million km (some 932,000 miles) from the Earth at a point where the gravity of the Earth and the Sun balance out. This means SOHO is able to keep the Sun under constant observation.

Since its launch, the SOHO craft has identified the main source of the solar wind as specific patches at the edges of the honey-comb shaped magnetic fields. There are large convection cells just below the surface of the Sun, each with an associated magnetic field. If we think of the cells as paving stones on a patio, then the solar wind breaks through like grass coming up through the cracks, says Helen Mason of Cambridge University. She also says that there is a difference in speed. Grasses don't grow at rates ranging from 8 km/s (5 miles per second) at the surface to 800 km/s (500 miles per second) of the solar wind!

The solar wind interacts with the Earth's magnetic field, changing its shape, which can damage satellites and disrupt communications and power systems. The world has increased its reliance on satellites, as they bring us our communications, our GPS systems and much more. Satellites like SAMPEX, POLAR, GOES and GPS are all used by scientists studying the solar wind.

The solar wind can vary in speed, but it can also vary in intensity. In May 1999, the solar wind density, the number of energetic electrons and protons per cubic centimetre, dropped to about 2% of its normal level, and stayed down for some 27 hours. The solar wind is an ionised gas, made up of charged particles, and the magnetic field shapes their flow, which means that the particles also shape the magnetic field that surrounds our planet. It's an action–reaction thing.

The speed of the remaining particles halved, and with so few particles pushing more slowly on the planet's magnetosphere, it blew up like a balloon to more than 100 times its usual volume.

The usual impact involves between 100 and 1000 million ions per square centimetre per second, each ion with an energy of at least 15 electron volts. If that all hit the Earth, a one degree by one degree square at the equator would be struck by 1 gram of material each second, about 57 grams (2 ounces) a minute, adding to the weight of the Earth.

Little of it gets through though. In actual fact, the solar wind, like desert winds, causes erosion. Not of the rocks, but of our atmosphere. The power of the wind is about a millionth of the power of solar radiation, but the solar wind is effective at stripping away the gases that surround the Earth-like planets, Mercury, Venus, Earth and Mars. For some reason, perhaps to do with the Earth's magnetic field, the effect on our planet seems to be less than it is on Mars. Even so, we probably lose around 3 kg (6.6 lb) of atmosphere each second.

Still, breathe easy because at that rate, it will take 50 billion years to dispose of our atmosphere, and around 15 trillion years to get rid of the oceans as well, and who knows, something else may happen.

solar flares

Solar flares spout out from the Sun all the time, but at a distance, they are hard to detect, mainly because the Sun is so bright. Sometimes though, the flares are so massive that they do become apparent.

On 1 September 1859, two astronomers, Richard C. Carrington and Richard Hodgson, were independently observing sunspots, using filters to limit the solar brightness, when they each detected a massive flare.

A typical flare has the energy of a few million large hydrogen bombs, and it involves radiation right across the electromagnetic spectrum. In 1859, the known spectrum only extended from infrared, through visible light, and on to ultraviolet – the astronomers could only see the visible part, but it was enough.

The gamma rays, X-rays and other hard radiation travelled as quickly as the light, but a huge blast of protons crossed from the Sun at 8 million km/h (about 5 million mph) and slammed into the atmosphere, shredding the ozone layer, and setting off spectacular auroral displays, all over the world. In the United States, currents were set up in telegraph wires that caused phantom signals to be detected in telegraph offices.

This sort of damage will come again: a 1989 solar flare caused surges that knocked out a Quebec power grid, and as our technology becomes more complex, there will be more lines, more circuits, more satellites and perhaps even more computers that are at risk.

The next big flare will travel at the same speed, 8 million km/h (5 million mph), giving us about 19 hours to get ready. It won't be a lot of time.

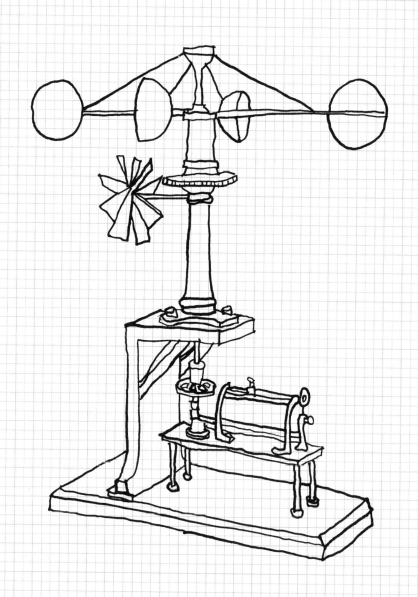

weather gauge

Life has existed on our planet for something close to 3.8 billion years. The jury is still out on the actual date, but it is at least 3.6 billion years, and there are some good hints that 3.8 billion may be a better bet.

The argument for 3.8 billion years is that very few of the oldest rocks are left on the surface of the planet, and there are definite fossils of well-developed life in the rocks dating to 3.6 billion years. That means there was probably life around before that. There are some older rocks which do contain traces of what might or might not be fossils.

Still, whenever the oldest life forms arose, they would have had something to talk about: the weather! The world was certainly hotter then, and even if there was no oxygen to breathe, way back before the algae started up photosynthesis, there were still gases to form gusts of wind, clouds to form lightning and rain, and maybe even the occasional bursts of hail, sleet or snow.

Throughout that time, weather is what has kept us going, bringing water, changing the soil and keeping the world fit for living in. Lightning for example makes a lot of the nitrogen compounds that plants and animals need to make proteins. We may grizzle about the weather, but without it, we wouldn't be here.

eddies in the air

No, not 'Eddie the Eagle' – whirly things. The earth whizzes around the Sun, but while we can see how the Sun appears to move through the field of 'fixed stars', we cannot really feel any of those effects, even though it is clipping along at almost 30 km/s (18.6 miles per second). The Earth, in turn, spins once every 24 hours, giving a linear speed at the equator of around 1600 km/h (1000 mph).

Even though this is faster than the speed of sound, we do not feel it because the air is moving at the same speed. We can detect the speed when the air at the equator moves north or south. Away from the equator, the linear speed of the planet is less, until at the poles, the linear speed is zero.

Figure 17. Fujita scale

Scale value	Speed (km/h)	Speed (m/s)	Speed (mph)	Description of damage
F-0	65–115	18–32	40–72	Light: tree branches broken, damage to chimneys and large signs.
F-1	119–180	33–50	74–112	Moderate: trees snapped, surface of roofs peeled off, windows broken.
F-2	184–252	51–70	114–157	Considerable: large trees uprooted, roofs torn off frame houses, mobile homes demolished.
F-3	256–331	71–92	159–206	Severe: roof and some walls torn off well-constructed houses, cars overturned.
F-4	335–418	93–116	208–260	Devastating: well-constructed houses levelled, cars and large objects thrown.
F-5	421–511	117–142	262–318	Incredible: strong frame houses lifted off foundations and destroyed, car-sized objects thrown more than 90 metres.

The air from the equator, when it drifts away, is going east faster than the ground beneath it, and if you were to look down from space, you would see cloud and wind systems curving clockwise in the northern hemisphere, and counter-clockwise south of the equator.

This is the Coriolis Effect, made famous in folklore because of the myth of the plughole. Basins, baths and toilets are all too small to be influenced by this effect, but large air patterns are not, and that is what gives us the twists and eddies which on the large scale are hurricanes, typhoons or cyclones. Each curls round a low pressure zone, clockwise south of the equator, the other way to the north.

On a smaller scale, there are tornadoes, small eddies rushing around a central low. The name 'tornado' is a corruption of the Spanish word for a thunderstorm, *tronada*, but the main damage is actually caused by the winds. The front of a tornado can move along at about 50 km/h (31 mph), but the winds, as described in the Fujita scale (Figure 17), go *much* faster. Measuring winds at those speeds can be hard, because people are busy staying alive. You can always look at the damage afterwards to gauge the wind speed, in the same way that earthquakes can be assessed afterwards on the Modified Mercalli Scale.

It also bears comparison with the Beaufort scales, which appear in Figure 18.

the beaufort wind scales

The Beaufort scale was developed for use by sailors in rough conditions when there might be no time to read the primitive instruments that were available. It is an empirical scale, developed in 1805 by Sir Francis Beaufort. It has been modified a number of times: with the age of steam, descriptions of sail behaviour were dropped, but it is still a handy reference. This table has been augmented with rough metric conversions, but wind forces 13 to 17, added to the scale in the 1940s, have been omitted. The land version is in Figure 18 and the scale on water is listed in Figure 19.

greased lightning

When you see lightning, try counting the seconds until you hear the thunder that accompanies it – this will give you an idea of how far away the storm is. It is fairly easy to count 'a thousand and ONE, a thousand and TWO, a thousand and THREE ...', and if you do this purposefully, you will be reasonably close to an accurate number of seconds. As a rough guide, three seconds is a kilometre and five seconds is a mile.

We can do this because the speed of light is much faster than the speed of sound, so we see the lightning flash almost as it appears, but we only hear the sound when the compression wave caused by local heating reaches our ears.

But how fast is lightning? If you answered that without thinking, you probably said that lightning travels at the speed of light: the names are similar, lightning produces light, and so that answer just sort of seems right.

Wrong! For starters, the lightning flash involves the travel of electrons, which have mass. If they travelled at the speed of light, they would have infinite mass. Looking at it another way, the air that carries the charge has resistance, just like any other conductor, and this slows the lightning bolt.

The best estimate is that the lightning bolt never travels faster than half the speed of light, and generally travels a great deal slower. At the beginning, the discharge travels around 30 m (100 ft) in a microsecond,

Figure 18. Beaufort wind scale on land

Force	km/h	m/s	mph	knots	Description	Land conditions
0	0–2	0–0.5	0–1	0–1	Calm	Calm; smoke rises vertically.
1	3–5	0.6–2	2–3	2–3	Light air	Direction of wind shown by smoke drift, but not by wind vanes.
2	6–11	3	4–7	4–6	Light breeze	Wind felt on face; leaves rustle; ordinary vanes moved by wind.
3	12–19	4–5	8-12	7-10	Gentle breeze	Leaves and small twigs in constant motion; wind extends light flag.
4	20–29	6–8	13–18	11–16	Moderate breeze	Raises dust and loose paper; small branches are moved.
5	30–39	9–10	19–24	17–21	Fresh breeze	Small trees in leaf begin to sway; crested wavelets form on inland waters.
6	40–50	11–13	25–31	22–27	Strong breeze	Large branches in motion; whistling heard in telegraph wires; umbrellas used with difficulty.
7	51–62	14–17	32–38	28–33	Near gale	Whole trees in motion; inconvenience felt when walking against the wind.
8	63–75	18–21	39–46	34–40	Gale	Breaks twigs off trees; generally impedes progress.
9	76–87	22–24	47–54	41–47	Severe gale	Slight structural damage occurs (chimney-pots and slates removed).

Figure 18. Beaufort wind scale on land (continued)

Force	km/h	m/s	mph	knots	Description	Land conditions
10	88–102	25–28	55–63	48–55	Storm	Seldom experienced inland; trees uprooted; considerable structural damage occurs.
11	103–117	29–32	64–72	56–63	Violent storm	Very rarely experienced; accompanied by wide-spread damage.
12	118–132	33–37	73–83	64–71	Hurricane	

Figure 19. Beaufort wind scale on water

Force	km/h	m/s	mph	knots	Description	Land conditions
0	0–2	0–0.5	0–1	0–1	Calm	Sea like a mirror.
1	3–6	0.6–2	2–3	2–3	Light air	Ripples with the appearance of scales are formed, but without foam crests.
2	7–11	2–3	4–7	4–6	Light breeze	Small wavelets, still short, but more pronounced. Crests have a glassy appearance and do not break.
3	12–19	4–5	8–12	7–10	Gentle breeze	Large wavelets. Crests begin to break. Foam of glassy appearance. Perhaps scattered white horses.
4	20–29	6 8	13 18	11–16	Moderate breeze	Small waves, becoming larger; fairly frequent white horses.

Figure 19. Beaufort wind scale on water (continued)

Force	km/h	m/s	mph	knots	Description	Land conditions
5	30–39	9–10	19–24	17–21	Fresh breeze	Moderate waves, taking a more pronounced long form; many white horses are formed. Chance of some spray.
6	40–50	11–13	25–31	22–27	Strong breeze	Large waves begin to form; the white foam crests are more extensive everywhere. Probably some spray.
7	51–62	14–17	32–38	28–33	Near gale	Sea heaps up and white foam from breaking waves begins to be blown in streaks along the direction of the wind.
8	63–75	18–21	39–46	34–40	Gale	Moderately high waves of greater length; edges of crests begin to break into spindrift. The foam is blown in well-marked streaks along the direction of the wind.
9	76–87	22–24	47–54	41–47	Severe gale	High waves. Dense streaks of foam along the direction of the wind. Crests of waves begin to topple, tumble and roll over. Spray may affect visibility.

Figure 19. Beaufort wind scale on water (continued)

Force	km/h	m/s	mph	knots	Description	Land conditions
10	88–102	25–28	55–63	48–55	Storm	Very high waves with long over-hanging crests. The resulting foam, in great patches, is blown in dense white streaks along the direction of the wind. On the whole the surface of the sea takes on a white appearance. The 'tumbling' of the sea becomes heavy and shock-like. Visibility affected.
11	103–119	29–32	64–72	56–63	Violent storm	Exceptionally high waves (small and medium-sized ships might be for a time lost to view behind the waves). The sea is completely covered with long white patches of foam lying along the direction of the wind. Everywhere the edges of the wave crests are blown into froth. Visibility affected.
12	120–132	33–37	73–83	64–71	Hurricane	The air is filled with foam and spray. Sea completely white with driving spray; visibility very seriously affected.

about $^1/_{10}$ of the speed of light, but then there is a pause of perhaps 50 microseconds before the next step. In the end, the whole discharge may continue for several milliseconds, and sometimes there will be several extra flows of charge up and down the same column. In short, there is no easy, accurate answer, but at least we can say that 'at the speed of light' is an inaccurate answer.

groundwater

In the early 1860s, there was a race on in Australia, with different colonial capitals desperately seeking a route that would bring the telegraph line from Asia (and so from Europe) to their capital first – knowledge was power.

Famously, several members of the ill-fated Burke and Wills expedition (including both Burke and Wills) died of starvation, scurvy, thirst and stress in search of such a route. Far to the west, a wild and canny Scot, John McDouall Stuart, made repeated assaults through central Australia and eventually succeeded in finding a usable route. More importantly though, it was a route which had water, essential in the heart of a brutally arid land.

By a combination of luck and good management, Stuart worked his way along the edge of the Great Artesian Basin, a monster underground lake that lies under 22% of Australia. It draws in water from way off in the lush tropics of coastal Queensland, seeping through porous rocks that are protected from the drying sun by a massive overburden of rocks, sand and dust. The estimated speed for a given molecule is between 1 and 5 m (3.3 and 16.5 ft) per year.

At the end of its journey, the water emerges, carrying dissolved salts that crystallise out to form characteristic piles called mound springs.

It was these springs which made Stuart's telegraph route work, because they gave water to the operators in the repeater stations dotted through central Australia.

The water in a good aquifer may flow 125 m (410 ft) a day under the best conditions, 15 m (49 ft) a day under average conditions. Because the flows occur in such a broad 'pipe' in the Great Artesian Basin, there is little need to worry about how fast the water flows: the big worry is how fast it is being lost. Only in recent years have people realised that these losses are happening too fast.

a flow of ice

There is a hard-to-kill legend that glass is a liquid and that panes of glass in medieval cathedrals have sagged slightly, and are thicker at the bottom than at the top. It is true that glass is a non-crystalline solid, so it could reasonably be described as a 'liquid' from the way its molecules are mixed, *but glass does not flow*. Not ever, not even over a few centuries.

Ice is a different matter altogether. Ice flows when it is in a big enough mass, and it flows in a very characteristic way. It oozes its way down a valley, slowly and majestically, while down on the valley floor, loose boulders and bits of gouged-off stone rip at the rock, scraping it smooth and leaving the valley with a characteristic U-shape that can be seen long after this flowing ice, the glacier has gone.

Glacier flow depends on the slope and the rate at which snow accumulates and packs to form ice that presses on the ice below. In the end, all things being equal, the rate of flow just matches the rate at which new ice is acquired, but this is weather that we are discussing, and nothing is ever equal. Ice and snow collect faster in winter, and summer warmth can melt the front face of the glacier, so the wall of ice at the glacier's end appears to retreat, only to advance again in winter.

During an Ice Age, the glaciers advance, new glaciers form where there were no glaciers (even at the top of Hawaii's Mauna Kea), and existing glaciers may merge with others and flow even further. To us, glaciers may seem like threatening and evil things, but to those living in dry country below a glacier, the slow but steady summer flow of meltwater is a lifesaver.

In India, 600 million people depend on rivers which are fed by Himalayan glaciers. Sadly, a number of rivers crossing the north Indian plains could run dry within 50 years, given the speed at which Himalayan glaciers are retreating. A number of smaller Himalayan and Kharakoram glaciers have already disappeared entirely. The Khatling glacier, one of those which used to feed the Ganges, is still marked on trekking maps, but is now just an empty valley, free of ice.

Gangotri, the biggest glacier in the Garhwal Himalayas, the mountains which are the main water source for the Ganges, has a face which is receding up into the mountains at almost 1 km (0.6 mile) a year. In time, there will be no glacier left to deliver summer water.

The glaciers really are disappearing. There were 27 glaciers in Spain in 1980, but by 1998, that number had dropped to just 13, and 10 by 2000. Glaciers in Spain covered 1779 hectares in 1894, down to 290 hectares in 2000, with most of the change having happened since 1980.

All over the world there is a clear trend in glaciers, though that trend is not universal. One recent study of more than 300 glaciers over 1986–2002 showed 102 more or less standing still, 40 advancing (a gain of 7.1 km^2 (2.74 miles^2) in area) and more than half, 171, retreating, for a loss of 146.1 km^2 (56.41 miles^2).

Many of the non-retreating glaciers are tidewater glaciers, which end at the sea, but these seem to reach a critical point and then fall back fast. The world's fastest glacier, the Columbia, increased its speed of forward flow from 25 to 35 m (82 to 115 ft) per day in 1999 and by 2005, it was still retreating fast, thinning even as it continues to rush forward, calving icebergs and melting faster than it advances.

Some people and a curiously large number of qualified scientists say that the retreat of the glaciers is a warning. Others say that global warming is a myth. You choose!

agassiz measures the glacier

While the Columbia glacier of Alaska moves 25 m (82 ft) a day or more, most glaciers, especially those within easy reach of inhabited areas, are far slower. That means that unless people looked very carefully, it would be hard to detect any movement at all.

After becoming curious about glaciers, Louis Agassiz, a Swiss, established a camp in 1840, using a huge boulder and a blanket to form a summer shelter. He had hoped to use an old hut that he had seen the previous year, but he found the remains of the hut, crushed and scattered, some 60 m (200 ft) down the glacier from where it had last been seen. He exclaimed 'aha!'

Towards the end of the season, Agassiz bored a series of holes in a line across the surface of the glacier and carried up a set of stakes which he inserted into the holes, but there is no record of what he saw of those stakes later.

By this time, Agassiz had learned to recognise the tell-tale signs of glaciation: the moraines where a glacier had dropped the stones it carried, the U-shaped valleys and such, but few scientists accepted his theory that the world had been shaped by ice ages, so he travelled to Scotland, and sought out examples there. Not many of those he spoke to in Scotland accepted his ice-age notions, but those who did were influential, and his ideas were slowly accepted.

In September 1841, Agassiz drove a new set of stakes into the ice, which showed the next spring that the centre of a glacier flowed fastest. He added even more in the summer of 1842 from which he measured daily and nightly speeds, and more. With proof that glaciers moved and a measure of how fast, his ideas spread even faster.

icebergs ahead!

Off the north east coast of Newfoundland, icebergs drift at about 0.2 m/s (8 inches), or 0.7km/h (0.4 mph). The speed depends on currents, waves and wind, but with so much of the iceberg under water, wind and waves have a less marked effect. On occasions, speeds can get above 1 m/s (3.3 ft/s), 3.6 km/h (a little over 2 mph), but even that adds up. In the course of a day, an iceberg has the potential to move almost a degree of latitude and well over a degree of longitude in waters close to the polar regions where the meridians are crammed together.

In the Antarctic, if an iceberg is caught up in the Antarctic Circumpolar Current, it may take off at an astounding 7 km/h (4.3 mph), making tracking something of a challenge. Any penguin hitching a ride on such a lump of ice could swim back, but might be better advised to go with the floe.

The Gulf Stream operates at similar speeds, and at one time produced an extra hazard for shipping in the Atlantic by shifting icebergs from one place to another. There is more to the iceberg collision picture than the *Titanic*, which was neither the first nor the last, though that particular event did have a remarkably high loss of life.

Between 1686 and 2000, there were 586 recorded incidents where vessels were damaged in collisions with icebergs. Most of these were in waters in high latitudes, and the reasons why become apparent to any wielder of Google Earth, or any other internet system that lets you look at aerial photographs of the area off the coast of Greenland or Antarctica. Search for a bit and you will find a swarm of white dots which could be anywhere from 200 to 1000 km (124.3 to 621.4 miles) further off, just a week later. Even slow speed kills, under the right conditions.

raindrops and hail

It can be argued that a drop of water typically travels long distances, even thousands of kilometres (miles), before it falls to the Earth again as rain, sleet or snow. For most of that time, however, the drop is not a drop at all, but a mere collection of droplets and molecules, moving at the whim of air currents, lacking the mass to make any concerted plunge towards the ground.

A cloud is what forms when air cools, so the amount of energy needed to split up the mini-droplets is no longer there. The cloud we see is made up of groups of water molecules that are large enough to interact with light to reflect it towards us (white clouds) or away from us (black clouds).

Clouds are filled with air currents passing up and down, swirling the droplets around. Under the right conditions, droplets start to collide and join together and fall out of the cloud. Again, under the right conditions, the drops may be carried by updrafts back into the cloud, cooling as they go: whenever air rises, it moves to a region of lower pressure, so the gas expands, and it is a given of physics that expanding gas gets colder. Sometimes, hail or snow forms.

A large drop of rain, about 5 mm (0.2 inches) across, has a terminal velocity of around 9 m/s (29.5 ft/s), while 6 mm (0.24 inch) raindrops fall at 10 m/s (33 ft/s). These sorts of drops carry enough energy to cause damage to exposed soil. Drizzle, tiny drops less than 0.5 mm, float gently down at 2 m/s (6.5 ft/s), or 7 km/h (4 mph), settling and wetting the soil, rather than blasting and devastating it.

If the cold water becomes hail, it has a terminal velocity proportional to the square root of its diameter. A 1 cm (about $^1/_3$ inch) hailstone reaches 50 km/h (31 mph). The largest known hailstone, 144 mm (5.7 inches) in diameter, hit at 47 m/s (154 ft/s), or 169 km/h (105 mph).

quick thinking

Fast communication has long been regarded by humans as important. Kings and emperors have always expected that news will reach them fast, whether it relates to military threats or any other sort of event. Until the first semaphores in the late 1700s, fast news travelled with runners, or even faster news, with messengers on horses.

When Mr Huskisson, MP, died in a panic under the wheels of a train travelling at a mere 48 km/h (30 mph), he died because he had grown up in a world where the fastest speed anybody knew was the speed of a galloping horse. It was also a world in which energy was measured by the power of a horse, and many of us still assess an engine in terms of its horsepower.

The people who lived in the 300 years before Huskisson died had seen almost no changes in the speed of things. The people who saw out the next 30 years after Huskisson, learned to live with speed, to glory in it, and to realise that horses were no longer the fastest or most powerful things that they would see. In *A Midsummer Night's Dream*, Puck tells Oberon, 'I'll put a girdle round about the earth in forty minutes'.

In reality, a low-level satellite takes more like 90 minutes to circle the planet once, but information can now travel very much faster. We can also process information very much faster, but even if railroads were faster than horses, animals were more flexible, so it took a while for people to switch from the horse as a medium for transmitting messages to the horse as a means of transmitting money to the bookmakers.

horse messengers and pony expresses

On 4 July 1826, the 50th anniversary of the signing of the American Declaration of Independence, its main author, Thomas Jefferson, the third president of the United States, lay dying at Monticello, his home in Virginia. John Adams, the second president of the United States, died in Massachusetts a few hours later, noting as he died that 'Thomas Jefferson survives'.

With no telegraph, no radio, no railway system, communication was never fast, and any news travelled slowly.

By the year of the great solar flare, 1859, a third of a century later, most of the United States was connected by telegraph wires, and a message could be sent from New York to St Louis, Missouri and a reply could be sent back, all in one day. California was yet to be connected with St Joseph, Missouri – that didn't come until 1861. Until then, messages had to be carried by the Pony Express from one end of the closing gap to the other.

In Peru, for the Incas to communicate, with no written language, all messages had to be verbal, and they were carried by runners on foot. It was said that a message could go from Quito to Cuzco, 1980 km (1230 miles), in 5 days, and tests have shown that with trained runners, this was possible. The mounted couriers of the Romans were lucky to travel 160 km (100 miles) in a day but they probably stopped at night, while the runners on foot could continue on the roads constructed by the skilled Peruvian stonemasons.

Other nations also had courier systems. Egypt had them 4000 years ago, Babylon's Hammurabi had couriers in about 1750 BC (they covered 200 km (124 miles), in two days), and Nebuchadnezzar's couriers are even mentioned in the Old Testament, in Jeremiah 51:31: 'One post shall run to meet another, and one messenger to meet another, to show the king of Babylon that his city is taken ...'

Darius of Persia used chains of shouting men, positioned on high points, and Julius Caesar explained that the Gauls used a similar method to gather their warriors to arms. For most purposes though, men on horses were the fastest way of sending a message until about 200 years ago.

China was using courier relay systems by about 1000 BC, and using the Royal Persian Road, couriers could travel 2750 km (1708 miles) in seven days and nights. Using horses, this was a close comparison with what the Inca runners could achieve.

In one case, the system was based on research. Cyrus the Great (599–530 BC) experimented to see how far a horse could ride without breaking down when hard-ridden, and then had post stations erected at those distances. In all horse messenger services, the posts have been between 20 and 25 km (12.5 and 15.5 miles) apart, with riders changing horses at each station, and riding between four and six stages at around 16 km/h (10 mph).

At the end of their stages, riders were also worn out. The metabolic cost on the rider of hard horse riding is 1.9 to 2.4 litres (0.5 to 0.6 gallons) of oxygen per minute, 66.4% of maximum metabolic power. This can be maintained for an estimated 276 minutes (4 hours, 36 minutes), in which time the rider would have covered four or maybe five stations.

At the end of 1892, two-thirds of a century after Adams and Jefferson died, the whole world was connected by telegraph cables under the sea, and in 1926, a hundred years after they died, news could spread around the world by wireless telegraphy, which was already beginning to be known as radio. The world was linked to such an extent that even an economic recession could engulf the world in no time at all.

the coming of the telegraph

The word 'telegraph' was coined by Claude Chappé, who introduced his telegraph system (which we would today call a semaphore) to France in 1793, speeding up communications during the Napoleonic wars. This form of telegraph was quickly taken up by the English, and leaves its mark today in places called Telegraph Hill, where the stations were set up.

The method used was a relay system, with each operator reading a signal from one station, and repeating it to the next, until it reached the end of the line of stations, or its destination. The French system mainly followed the coastline, and a few of the stations, or their ruins, are still to be found around Brittany. In both Britain and France, the telegraphs were used mainly for military messages and orders, and served to make warfare more efficient.

They had their drawbacks, though, so the idea of an electrical telegraph always had support: a landing party only had to destroy one semaphore station to break the chain, and the semaphores could not work in heavy rain, fog or at night. There had actually been an electrical telegraph, using static electricity, in London in 1729, but the real heyday of the electric telegraph came in the 1830s and 1840s.

There was a curious feasibility study carried out near Paris in 1746, when the Abbé Jean-Antoine Nollet arranged a 1 mile (1.6 km) circumference circle of 200 Carthusian monks, each pair of monks linked by thick iron wires, described as '25 feet long' (about 8 m).

He then connected the two end monks to a Leyden jar, a primitive but effective capacitor or charge storage device. Nollet wanted to find out if the electric charge travelled instantaneously, and the experiment succeeded beyond his dreams, though probably giving the monks involved a few nightmares. Shortly afterwards, he repeated the demonstration, using Royal Guards in the presence of the King. Small wonder the French had a revolution!

After that, everybody knew that electricity was the way to go, but there were a few challenges: getting a good source of electrical power was first, solved when Alessandro Volta invented the pile, or battery, in 1800. Then there was the problem of insulating the wires so that they would carry a signal over a long enough distance, without all of the charge leaking into the earth. People began with cotton-covered wires, invented for making hats, but also tried enamelled wires.

They later discovered a rubbery substance called gutta percha, and then rubber itself. But before even that, there was another problem: even with good insulation, signals died out over distance. Some sort of gadget was needed to amplify the signal. In America, Joseph Henry invented the electric relay for this purpose, and Edward Davy invented it independently in Britain.

By now, people were trying out telegraphs in all sorts of places. William Cooke saw one demonstrated in Germany and combined with Charles Wheatstone to develop a British electrical telegraph. They bought Edward Davy's patent in order to get the rights to the relay, and a few years later, their telegraph was used to alert London police to the presence of a fleeing murderer on a train to London. The murderer thought speed would save him, but the telegraph was faster.

All the same, the Cooke-Wheatstone telegraph was by all accounts still slow, with swinging needles being pointed at a set of 20 letters. C, J, Q, U, X and Z were all left out. It was not the sort of telegraph you would use to send a quick quiz about jazz, but it worked well enough – that is, until the clever Mr Samuel F. B. Morse came along.

The Morse code has now been largely forgotten, but it combined long and short signals (dots and dashes, dits and dahs) to produce letters. Dit-dah was A, dah-dit-dit-dit was B, and so on, with shorter sequences for more common letters. Morse thought the system would be used by printing the dots and dashes on tapes that would later be transcribed by clerks, but at the beginning, an operator would read off the tape as it came through, dictating the letters to a second clerk before a third wrote a fair copy.

Before long, operators were receiving 'on the click', listening rather than watching, and writing down the letters they now heard, rather than saw. Just as beginning readers have to look at each letter and 'sound out' a word, the telegraphists began by listening to letters and improved. Before long, they could hear not only just the letters, but whole words in one grab, and messages could pass over the wires at 2000 words an hour.

By the 1860s, successful cables were being laid and planned all around the world, and some of the changes were surprising. Once, when steamers arrived in New York from England, or in Rio, Cape Town or Sydney, for that matter, journalists would leap into the boats and rush to greet the new arrivals, scrounging foreign papers to get the news, and to see what other events had happened in the world.

With telegraph cables linking the world, foreign papers were still welcome, but more as sources of extra information, too minor for the expensive telegraph cable. The world was changing fast.

Who would have predicted, back then, that a system based on the telegraph system and its later incarnation, the telephone, would become what we know now as the internet, delivering foreign papers to printers on the other side of the world, or to the screens of individual readers?

rubik's cube and lightning chess

In the birthplace of the Rubik's Cube, Budapest, in October 2007, world champions gathered to strut their stuff. Anssi Vanhala of Finland won in the feet-only challenge when he lined up the colours of the six-sided classic 3x3 cube in just 49.33 seconds. This was nothing when compared with the hands of Yu Nakajima, who achieved an average of 12.46 seconds as he reset the six different colours of a 3x3 cube.

Even that, though, was outside the world record of 9.86 seconds set in May 2007 by Frenchman Thibaut Jacquinot. The Budapest event was organised to celebrate the 25th anniversary of the introduction of the cube, which seems to have somewhat lost the cult status it held in the 1980s. It is possible that these records will never be eclipsed, but they said that once about the 4-minute mile ...

The sort of mind that can wrap itself around a Rubik's Cube is the sort of mind that likes chess. The Rubik's Cube winners and record holders, would probably be more attuned to the different variants of lightning chess, where players have limited time for all of their moves: 5 minutes for blitz, 3 for fast blitz and 1 minute for bullet chess.

the lightning calculators

Johann Carl Friedrich Gauss grew up to be a very clever mathematician, but as a boy, he suffered the misfortune of having a lazy teacher. According to a pretty legend, on one occasion, the teacher told the class to write all of the integers from 1 to 100 and add them, assuming that this would give him an hour of relaxation.

Gauss' hand went up immediately. The answer, he said, was 5050, which the teacher knew to be correct, as he had laboriously calculated it himself, many years earlier. Angry that the boy must have been told the answer by a previous student, he challenged him to prove it by showing how he had arrived at this value.

Gauss explained that the two outside numbers, 1 and 100, sum to 101, as do 2 and 99, 3 and 98, and so on. In all, there are 50 pairs, all summing to 101. Then, 100 x 50 is 5000, add 50, and there is your answer. Gauss continued to perform complex mental calculations later in life, and he was by no means the first or last of his kind.

In *The Maths Gene*, Keith Devlin describes seeing Arthur Benjamin give a demonstration of his ability to find the square roots of six-digit numbers. Before starting, Benjamin asked that the air conditioning in the room be turned off. While he and the audience waited for that to be done, he explained that the most important thing was to hear the numbers, and that the hum might interfere.

George Parker Bidder, born in England in 1806, also stressed how the sound of numbers was important when he was giving stage demonstrations of his skills of calculation. Memory probably also played a part in this case though. One of his brothers knew the whole of the Bible by heart, and another brother, after losing his books in a fire, rewrote them from memory, in the space of 6 months.

George von Neumann, one of the pioneers of computing, also had amazing calculating skills, and there are a number of interesting stories about him. There is a standard problem, where two trains start at opposite ends of a 200-mile track at 50 mph. A fast fly sets off from the front of one train at 75 mph, turns at the other train and flies back, repeating this until the trains collide and the fly is crushed. How far did the fly travel?

The answer, 150 miles, is easy to get at if you think cleverly. When he heard the question, von Neumann gave the answer straight away. The questioner said it was strange, that most people tried to sum an infinite series. 'What do you mean, strange?' asked von Neumann. 'That's how I did it!' It's said that von Neumann also had an amazing memory: he could read any book once, and afterwards, recall each word of it.

Some of the lightning calculators are unfairly called idiot savants, meaning people who just have the single skill of calculation. Just as a good musician can look at a musical score and 'hear' what is on the page, so some calculators can feel, see or hear the answer to a complex problem, although memory could play a part in it.

Once, when American Zerah Colburn was asked to square 4395, he hesitated for a moment, but when the question was repeated, he offered the correct answer. He explained that he did not like multiplying four-digit numbers together, but he had squared 293, then multiplied the result by 15, twice over. The number 4395 is, of course, the product of 15 times 293.

On occasion, mathematicians show a remarkable familiarity with numbers which have a specific meaning for them. Ordinary Britons, Americans and Australians would all recognise at least one of 1066, 1776 and 1788, but how would you respond to 1729? I once challenged a mathematician colleague, saying that my computer had a four-digit password that I believed he could get, first try. To the amazement of some colleagues, he did just that. A year earlier, we had found we both knew and enjoyed the tale of Ramanujan, the taxi and 1729, and as the password had to be a special number we both knew, he chose it.

Ramanujan was an instinctive mathematician, and the number 1729 is a special type of number called a Carmichael number. The English mathematician G. W. Hardy mentioned the number casually to Ramanujan as the number of a cab he had just travelled in, adding that the number was uninteresting. In all probability, Hardy was trying to find out if Ramanujan had instinctively and independently identified the Carmichael numbers, but he was unprepared for the Indian mathematician's answer.

'Oh no, Hardy, it is a very interesting number. It is the first number that is the sum of two cubes in two different ways!' This is correct: the number is the sum of 9^3 (729) and 10^3 (1000) and also of 1^3 (1) and 12^3 (1728). Ramanujan probably also knew that 1729 is the product of three primes which are in arithmetic progression, but he apparently did not find that as interesting as the sum-of-two-cubes solution.

Some of the people we call idiot savants only have the one skill: fast calculation, but they are by no means idiots. Others combine lightning calculation with high intellect. George von Neumann was one such 'idiot', and A. C. Aitken, a professor of mathematics at Edinburgh, was another. Aitken could recite the first 1000 digits of π (pi) from memory, and when confronted with the number 1961, he could break it down into the following: 37 x 53, or $44^2 + 5^2$, or even $40^2 + 19^2$. Wim Klein, another calculating wizard, is quoted as saying that for some people, 3844 was just 3-8-4-4, but he would react 'Hi, 62 squared!'.

Pocket calculators and computers have not done away with the mathematical prodigies, though they are less likely to find steady work on the stage. All the same, in 2005, a French student, Alexis Lemaire, used memorised tables to calculate the 13th root of a 200-digit number, in his head. In less than 9 minutes from when he was given the number, he provided the 16-digit answer: 2391481494636373. As Fermat would say, there is insufficient space in the margin of this book to write it down.

lightning cube roots

Occasionally, a stage act is based on a very small amount of memorisation and a lot of trickery, as in this example, which I created for this book. Suppose I tell you that I can find the cube roots of every perfect cube between 1 million and 8 million, and I give you a calculator, inviting you to enter a 3-digit number, less than 200, and then read out the result after multiplying it by itself twice.

You might choose to enter the value 173, which would provide you with the value 5,177,717. When you read this out from the calculator, I would immediately tell you that the number you had entered was 173.

The trick to this does not involve memorising all of the cubes up to 200^3. That would be possible, but it is unnecessary. You only need to memorise the cubes up to 20, because adding three zeroes will give you the cubes of all of the multiples of 10 up to 200. Note that you can extend this further if you wish, but for this example, that is plenty. My lower limit of 1 million means I can concentrate on 3-digit numbers. Again though, the trick can be extended. I start with a table of cubes, where I only need an approximate value:

number	100	110	120	130	140	150	160	170	180	190	200
cubed (millions)	1	1.3	1.7	2.1	2.7	3.3	4	4.9	5.8	6.8	8

Next we need to know a curious feature about the last digit of any perfect cube and how it relates to the cube itself:

number ends in	0	1	2	3	4	5	6	7	8	9
cube ends in	0	1	8	7	4	5	6	3	2	9

Now you have my secret: when you say 'five million, one hundred ...' I know we are between 170 and 180. Then I listen for the last digit of the cube (7), so I know the last number of the cube root is 3.

the meaning of the mulgawood mercury

Many numerical stage acts in the past have relied on the performer's ability to recall long strings of numbers. Amateurs may rely on a mnemonic like this: 'To express e, remember to memorize a sentence to simplify this.' To find the value of e, count the letters to retrieve 2.7182818284.

Figure 20. Consonant tricks

Digit	Consonants	Reminder to help you remember
1	T or D	T has one downstroke
2	N	N has two downstrokes
3	M	M looks like a 3 rotated 90° counter-clockwise
4	R	R is the fourth letter in FOUR
5	L	L is 50 in Roman numerals
6	J, soft G, SH, CH	J looks like a 6 reversed
7	K, hard G, hard C	K can be made from two 7s
8	F, V or PH	A lower-case script f has two loops, like 8
9	P, B	P looks like 9 when reversed
0	Z, S or soft C	Z is the initial of Zero

As a side note, I find a value of e to six significant digits sufficient, and my mobile phone number ends in 271828 — the friends I prefer think this is seriously neat, the rest can go hang themselves! The whole number is $[\sqrt{2}-1]$ e, if you follow me.

Professional stage performers rely mainly on a method developed by Pierre Hérigone in 1634. It begins with memorising a table of consonants like the one in Figure 20, until it is automatic to link each consonant with its specified value.

The trick is to make an association or an image to recall, using these consonants to find words that use them and no others. To take a simple example, oxygen has an atomic weight of 16, so DaSH or TouCH would be suitable reminders. Somebody who DaSHes needs a TouCH of oxygen, and you have your mnemonic.

The element indium (which chemists order in very large bottles so they can play carboys and indiums) has the atomic number 49, which gives us either the combination R-B- or R-P-. As we cast around: rubber, robber, rope, rupee, there we have it: the RuPee is the Indian unit of currency, which links neatly to indium.

The experts will often take shortcuts, so the square root of 6 is recalled without the leading 2 (people using tricks like this know that the value has to start with 2), so the standard mnemonic is RaRe Bee, where you recall that the standard honey-comb is made up of sets of hexagons. The value is 2.449.

Numerologists are always finding new things to recall. In Psalm 46 in the King James Bible, the 46th word is 'shake', while counting backwards, the word 46 from the end is 'spear'. And in 1610, when the new translation was first printed, William Shakespeare was 46! If you think that means anything at all, you would be very RaSH, but if you TouCH your ToeS, you will be able to remember the year in a flash.

If you get confused and TouCH your Nose, you will have the decimal part of the square root of 10 (3.162), so apply this trick with care.

Mercury boils at 357°C, MLK or MLG, so you might picture the Greek god Mercury MiLKing a cow or buzzing around on his winged sandals with a bottle of milk. In Australia, tacky *objets d'art* were carved from the wood of *Acacia aneura*, or to give it its Australian name, mulga. For me, that value will always live in a rendition of the lively god in dark brown mulgawood.

fibonacci's serious rabbits

The early 1200s were very much the Middle Ages, but even then, steps were being taken that would lead, in time, to the discoveries of the Renaissance. Most of those steps involved advancements in trade, which led to new goods being introduced to new places, and along with them, new ideas.

Young Leonardo of Pisa was the son of a merchant who carried the nickname 'Bonacci', which you can take to mean either 'good-natured' or 'simple', according to taste. I prefer to believe that Bonacci, like his son, was far from simple. Anyhow, Bonacci ran an Italian trading post in what is now Algeria, and he produced Leonardo, a very bright son, who joined him there in the late 1100s. While there the boy picked up on the Arabic (or Hindu) system of writing numbers.

Leonardo knew a good thing when he saw it, so he travelled around the Mediterranean, studying with various Arab scholars. In 1202, he published his *Liber Abaci*. Literally 'the book of the abacus', this work introduced this new counting system in terms that tradesmen and academics could both understand, giving practical examples.

It was by no means the first book to mention Hindu-Arabic notation, but it took off. This was probably due, at least in part, to the practical examples he provided. In one, he examined the way rabbits breed like, well, rabbits. Everybody knew how fast the rabbit numbers grew, but what was the mathematics behind it?

He assumed that rabbits take a month to mature, that they breed and produce two young, a male and a female after one more month, and that rabbits live for 12 months. With that assumption, he wondered how many rabbits there would be at the end of this time, starting with two newborn rabbits.

At the end of the first month, the rabbits mate, and there is still just one pair. At the end of the second month, the doe gives birth and there are two pairs. The parents mate again, and at the end of the third month, there is a third pair. The first of the new young and the parents both mate and produce young at the end of the month. Now there are five pairs, three ready to breed, and two immature. At the end of the next month, there are eight pairs, and so on. The total number of rabbit pairs, taking the start as Month 0 increases like this:

0	1	2	3	4	5	6	7	8	9	10	11	12
1	1	2	3	5	8	13	21	34	55	89	144	233

The point of this small puzzle was to show how much easier and quicker it was to add 89 and 144 than to add LXXXIX and CXLIV, the same numbers in Roman numerals. It caught on with the mob.

It did no harm that there is a somewhat mystic number, Φ (phi), known as the Golden Ratio, or the Golden Mean, defined by the simple equation $\Phi-1 = 1/\Phi$. This turns up in art, in Greek architecture, even in the proportions of normal pages, and when you divide any term in Leonardo's series by the previous one, you get values that get closer and closer to Φ, about 1.618. That really impressed people.

After his father died, Leonardo came to be known as Bonacci's boy, 'filius Bonacci' in mediaeval Italian, or Fibonacci for short, and that is why the extended sequence of numbers is known today as the Fibonacci series. Of course, if you wanted a continuing series, you needed to assume that the rabbits were immortal, but given his other assumptions about the rabbits, what was wrong with that? Then again though, maybe we should celebrate Leonardo, the Golden Mean, and his randy, speedy, rabbits by calling it the Phibonacci series.

the art of estimation

To a physicist, the notion of an immortal rabbit is quite acceptable. My English teacher encouraged me to psychoanalyse Macbeth, even though I objected that we shouldn't, since Freud hadn't been invented when Shakespeare was writing. Ever an historically minded cuss, I argued that it would be more relevant to look at the political situation in London, with a Scot sitting on the throne. Exasperated, he exhorted the class to engage in the willing suspension of disbelief.

And of course he did, if he wanted us to accept some of the artifices and conceits of coincidence found in the nineteenth century novel, but we scientific types were subjected to much hardier fictional nonsense than that.

We routinely solved problems that involved a steel girder of negligible mass, suspended at its centre of gravity by a silken thread. And before we were too far advanced, we heard our first physics joke. It was about three scientists, a mathematician, a geneticist and a physicist, who were trying to pick the winner of Australia's premier horse race, the Melbourne Cup, held each November.

The mathematician gathers a wealth of data on weather, rainfall, wind, pollen counts and other possible influences, and three years in a row, fails dismally to pick a winner. At the end of those three years, the geneticist has just finished drafting a plan for a breeding program that should, in five generations, produce a winner. And the physicist had got it right, three times in a row.

The others ask him how he did it. He reaches into his pocket and produces an envelope which he turns over. Then he draws a circle on it. 'Consider,' he says, 'a spherical horse running in a vacuum . . .'

In fact a spherical cow or a spherical horse can be a useful starting point to explore ideas and to get a first approximation that can be extended. Take the yarn about the bumblebee that was shown not to be able to fly: it's usually trotted out as evidence that scientists are thick, but there is a little more to it than that. In 1934, a French entomologist called Antoine Magnan tried to apply an engineer's equation to bumblebees.

He showed that according to that equation, designed for an aircraft that did not flap its wings, the bee could not generate enough lift.

There is a great deal of folklore wrapped around this 'event' and who was actually involved, but it appears that the equation was worked out by André Saint-Lagué, and while the incident is often dressed up as 'a scientist proving that bumblebees can't fly', all that was really shown was that the equation was inadequate to describe the flight of the bumblebee.

Magnan had shown that you can't apply *that* particular equation to bumblebees, rather like proving that spherical bumblebees can't fly, even if real ones, flapping their wings at 130 times a second, move happily along at 3 m/s (9.8 ft/s), 11 km/h (7 mph). Magnan's calculations merely showed that there was a faulty assumption in there somewhere, that the mathematical model was flawed.

As a child, when we escaped from the English classroom to the lab, we learnt of marvels that could be achieved with very simple apparatus. The muzzle velocity of a bullet could be measured with nothing more than a block of wood, a piece of string, a protractor and a measuring tape.

Our physics teacher, as equally at home with fiction as our English teacher, explained how, in the days of gunpowder and muzzle-loading firearms, slight variations in the ingredients, their amounts and proportions, could make a lot of difference. The most obvious measure was the speed at which a cannon ball or musket ball left the barrel of the gun, or in physics-speak, the muzzle velocity.

The idea was quite simple. You suspend a large block of wood and fire a bullet at it from close range. The bullet lodges in the block, and the energy of the bullet is transferred to the block, which swings like a pendulum. One simply has to measure the swing angle and then calculate the height the block reaches.

This device even has a name: it is called the ballistic pendulum, and it has been around since 1742, when it was invented by Benjamin Robins. From the swing, or so we were told, it is a fairly elementary calculation to estimate the energy and hence the velocity of the bullet. Unfortunately, this explanation ignores the 800-pound spherical horse rolling around the room.

Some of the energy goes into deforming the bullet and the wood, some is wasted as friction, and to do any calculations, we have to assume that the bullet stops instantaneously (which is as likely as a girder with negligible mass). Of course, if you are simply trying to compare different grades of gunpowder, rather than measuring the muzzle velocities, the losses will be similar in each case, and can be ignored. Whichever powder produces the biggest swing is the best, if everything else is kept constant — and in fairy physics, that always applies.

Robins was born to Quaker parents, but as a mathematician, tried to make gunnery a science. Along the way, his ballistic pendulum probably showed that Indian saltpetre made the best gunpowder. He died in India in 1751, supervising the construction of forts, and a few years later, the British drove the French out of India, which let them have all that excellent saltpetre for their own use.

fast books, slow books

How could one write *The Speed of Nearly Everything*, and not celebrate fast books? Or a few slow books as well really, because the word is usually a slow medium?

A fast book is usually written to meet an interest in a 'news event'. Biographies of politicians newly elevated to high office, or starlets given to low deeds, true crimes and the background to disaster – all of these tend to be fast books. Any potential for overlap between crimes, politics, starlets and disasters will not be considered here. Perish the thought that there might even be any link!

There are artificial fast books and compilations created as stunts, and sometimes those two categories overlap. In 1994, the work *100 Recipes in No Time at All* was generated (one can hardly put it more kindly) in 48 hours for the *Challenge Anneka TV Show*, a BBC series featuring Anneka Rice, where the format of the weekly show challenged the host to do something in a very short period of time.

In 2001, the BBC organised for the same editor of that 1994 work to put together a compilation of text messages sent in by viewers. As a social history document, recording texting usages and habits of 2001, that book might have some minor merit, but it was never going to be regarded as real literature.

A 2003 attempt by a consortium of 40 German writers set out to produce 'a book in 12 hours'. This time included the writing, editing, printing 1000 copies on a hand-press dating almost back to print pioneer Johann Gutenberg, and the binding. This was undoubtedly a stunt, but a stunt with good literary intentions, sponsored by *Stiftung Lesen* (the Reading Foundation) in Mainz, Gutenberg's birthplace.

Legend has it that Barbara Cartland could knock out a 'novel' in 10 days, assisted by a small horde of typists working from her tape-recordings. It is oft-repeated, but believable from literary analysis. Isaac Asimov wrote close to 500 works, if you include anthologies in which he appeared (estimates vary a bit, but it came close to 500). Asimov was successful and had access to assistants, and a few cynics claimed that he was, in fact, a committee. If so, it was a committee which stayed remarkably consistent over the years.

John Creasey published 562 books, all written by longhand, generally with five or six revisions. This includes the works he wrote under pseudonyms. He died aged 65 in 1973, having been a published author for 42 years (his first book was in 1932, and then the floodgates opened, and he averaged more than 13 books a year!).

Georges Simenon published for over 51 years, from 1921 to 1972, and is credited with some 450 works: including 200 novels and 150 novellas. He wrote '60 to 80 pages per day', probably equivalent to a Cartland-sized novel every six days. Charles Dickens would belt out a chapter a month of whatever novel he was working on, because most of them were serialised in magazines. I suspect he sometimes had more than one on the go at a time.

At the other end of the scale, James Joyce took 17 years to complete *Finnegans Wake* but the award for slowest books surely goes to those mute inglorious Miltons that never made it into print at all. Among those who were published, Margaret Mitchell's only published book was *Gone With the Wind*, and Harper Lee seems only to have published *To Kill a Mockingbird*. Helen Hooven Santmyer took 50 years to write *And Ladies of the Club* which was published when she was 88. Books don't come much slower than that.

index

C

D

E

F

First published in 2008 by Pier 9, an imprint of Murdoch Books Pty Limited

Murdoch Books Australia
Pier 8/9
23 Hickson Road
Millers Point NSW 2000
Phone: +61 (0) 2 8220 2000
Fax: +61 (0) 2 8220 2558
www.murdochbooks.com.au

Murdoch Books UK Limited
Erico House, 6th Floor
93–99 Upper Richmond Road
Putney, London SW15 2TG
Phone: +44 (0) 20 8785 5995
Fax: +44 (0) 20 8785 5985
www.murdochbooks.co.uk

Chief Executive: Juliet Rogers
Publishing Director: Kay Scarlett

Commissioning editor: Diana Hill
Project manager: Diana Hill and
    Paul O'Beirne
Editor: Elizabeth Anglin
Concept and design: Hugh Ford
Illustrations: Hugh Ford
Production: Kita George

National Library of Australia
Cataloguing-in-Publication Data

Macinnis, Peter.
The speed of nearly everything: from
tobogganing penguins to spinning
neutron stars / author, Peter Macinnis.
Sydney: Murdoch Books, 2008.
ISBN 9781741961362 (pbk.).
Includes index.
Speed—Miscellanea. Speed records.
531.112

A catalogue record for this book is
available from the British Library.

Printed by i-Book Printing Ltd in 2008.
PRINTED IN CHINA.